Building Community Disaster Resilience Through Private–Public Collaboration

Committee on Private–Public Sector Collaboration to Enhance Community Disaster Resilience

Geographical Sciences Committee

Board on Earth Sciences and Resources

Division on Earth and Life Studies

NATIONAL RESEARCH COUNCIL
OF THE NATIONAL ACADEMIES

THE NATIONAL ACADEMIES PRESS
Washington, D.C.
www.nap.edu

THE NATIONAL ACADEMIES PRESS • 500 Fifth Street, N.W. • Washington, DC 20001

NOTICE: The project that is the subject of this report was approved by the Governing Board of the National Research Council, whose members are drawn from the councils of the National Academy of Sciences, the National Academy of Engineering, and the Institute of Medicine. The members of the committee responsible for the report were chosen for their special competences and with regard for appropriate balance.

This study was supported by the U.S. Department of Homeland Security under Award No. HSHQDC-08-C-00176. Any opinions, findings, and conclusions or recommendations contained in this document are those of the authors and do not necessarily reflect the views of the agencies that provided support for the project. Mention of trade names, commercial products, or organizations does not constitute their endorsement by the sponsoring agencies.

International Standard Book Number-13: 978-0-309-16263-0
International Standard Book Number-10: 0-309-16263-7

Cover: Cover design by Francesca Moghari

Additional copies of this report are available from the National Academies Press, 500 Fifth Street, N.W., Lockbox 285, Washington, DC 20055; (800) 624-6242 or (202) 334-3313 (in the Washington metropolitan area); Internet http://www.nap.edu.

Printed in the United States of America.

THE NATIONAL ACADEMIES

Advisers to the Nation on Science, Engineering, and Medicine

The **National Academy of Sciences** is a private, nonprofit, self-perpetuating society of distinguished scholars engaged in scientific and engineering research, dedicated to the furtherance of science and technology and to their use for the general welfare. Upon the authority of the charter granted to it by the Congress in 1863, the Academy has a mandate that requires it to advise the federal government on scientific and technical matters. Dr. Ralph J. Cicerone is president of the National Academy of Sciences.

The **National Academy of Engineering** was established in 1964, under the charter of the National Academy of Sciences, as a parallel organization of outstanding engineers. It is autonomous in its administration and in the selection of its members, sharing with the National Academy of Sciences the responsibility for advising the federal government. The National Academy of Engineering also sponsors engineering programs aimed at meeting national needs, encourages education and research, and recognizes the superior achievements of engineers. Dr. Charles M. Vest is president of the National Academy of Engineering.

The **Institute of Medicine** was established in 1970 by the National Academy of Sciences to secure the services of eminent members of appropriate professions in the examination of policy matters pertaining to the health of the public. The Institute acts under the responsibility given to the National Academy of Sciences by its congressional charter to be an adviser to the federal government and, upon its own initiative, to identify issues of medical care, research, and education. Dr. Harvey V. Fineberg is president of the Institute of Medicine.

The **National Research Council** was organized by the National Academy of Sciences in 1916 to associate the broad community of science and technology with the Academy's purposes of furthering knowledge and advising the federal government. Functioning in accordance with general policies determined by the Academy, the Council has become the principal operating agency of both the National Academy of Sciences and the National Academy of Engineering in providing services to the government, the public, and the scientific and engineering communities. The Council is administered jointly by both Academies and the Institute of Medicine. Dr. Ralph J. Cicerone and Dr. Charles M. Vest are chair and vice chair, respectively, of the National Research Council.

www.national-academies.org

COMMITTEE ON PRIVATE–PUBLIC SECTOR COLLABORATION TO ENHANCE COMMUNITY DISASTER RESILIENCE

WILLIAM H. HOOKE, *Chair*, American Meteorological Society, Washington, DC
ARRIETTA CHAKOS, Urban Resilience Policy, Berkeley, California
ANN-MARGARET ESNARD, Florida Atlantic University, Fort Lauderdale
JOHN R. HARRALD, Virginia Polytechnic Institute and State University, Alexandria
LYNNE KIDDER, Center for Excellence in Disaster Management and Humanitarian Assistance, Washington, DC
MICHAEL T. LESNICK, Meridian Institute, Washington, DC
INÉS PEARCE, Pearce Global Partners, Inc., Los Angeles, California
RANDOLPH H. ROWEL, Morgan State University, Baltimore, Maryland
KATHLEEN J. TIERNEY, University of Colorado, Boulder
BRENT H. WOODWORTH, Los Angeles Emergency Preparedness Foundation, California

National Research Council Staff

SAMMANTHA L. MAGSINO, Study Director (from July 2009)
CAETLIN M. OFIESH, Study Director (until July 2009)
COURTNEY R. GIBBS, Program Associate
JASON R. ORTEGO, Research Associate (from November 2009)
NICHOLAS D. ROGERS, Research Associate (until November 2009)
TONYA E. FONG YEE, Senior Program Assistant (until September 2010)

GEOGRAPHICAL SCIENCES COMMITTEE

WILLIAM L. GRAF, *Chair*, University of South Carolina, Columbia
WILLIAM E. EASTERLING III, Pennsylvania State University, University Park
CAROL P. HARDEN, University of Tennessee, Knoxville
JOHN A. KELMELIS, Pennsylvania State University, University Park
AMY L. LUERS, Google, Inc., Mountain View, California
GLEN M. MACDONALD, University of California at Los Angeles
PATRICIA MCDOWELL, University of Oregon, Eugene
SUSANNE C. MOSER, Susanne Moser Research & Consulting, Santa Cruz, California
THOMAS M. PARRIS, ISciences, LLC, Burlington, Vermont
DAVID R. RAIN, George Washington University, Washington, DC
KAREN C. SETO, Yale University, New Haven

National Research Council Staff

MARK D. LANGE, Associate Program Officer
JASON R. ORTEGO, Research Associate
CHANDA IJAMES, Program Assistant

This report is dedicated to the memory of **Frank Reddish**, a long-time leader in natural disaster and recovery. Through years of committed and focused effort, Mr. Reddish made Miami-Dade County and the state of Florida a safer and more resilient place to live. His work drew attention and had impact both locally and nationwide. He contributed powerfully to this committee's information-gathering workshop, held September 9–10, 2009, and his work will continue to have a positive impact for years to come.

Preface

Recent national and international experience with natural and human-caused disasters highlights several realities. First, the planet on which we live—the planet on which we aspire to forge careers, establish marriages and families, grow economies, and seek peace and security—provides frequent and often unpredictable extreme events. Severe heat waves, cold snaps, and cycles of flood and drought determine what we call climate. Movement in the Earth's crust is manifested by earthquakes and volcanic eruptions. Environmental degradation, habitat loss, and reduction in biodiversity can occur incrementally but also through sudden devastation, such as through wildfire or an oil spill.

Second, extremes often trigger disruptions of communities that persist after the event that exceed a community's ability to recover on its own. These disasters are as much the result of human decisions as of nature. Land use, building codes, the engineering of critical infrastructure, distribution of wealth and poverty, and many other social decisions and actions shape the impacts of extremes and subsequent recovery.

Third, resilience to disasters is built at the community level. No community is immune to disasters, and no community is an island unto itself. The emerging role of critical infrastructure, just-in-time manufacturing, and the globalization of the economy means that all individuals and communities are interdependent.

Fourth, responsibility for building community resilience cannot rest with the public sector alone. In the United States, the public sector represents just ten percent of the workforce. The other ninety percent resides in the private sector—ranging from small, individually owned businesses to national and global enterprises—and in a range of non-governmental bodies and faith-based organizations. Operation and maintenance of many community assets, including critical infrastructure, remain in private hands. All sectors must collaborate to build community-level disaster resilience.

This report addresses these realities. It surveys what we know about effective private–public collaboration and how it may enhance community disaster resilience. It delineates areas where resilience-focused collaboration could benefit with more knowledge, and it lays out a comprehensive research agenda. However, the members of this committee note that in the face of rapid social change and technological advancement, our understanding of resilience–focused private-public sector collaboration is nascent. This report should be considered an initial exploration of a developing subject—not the final, definitive word.

William Hooke, *Chair*
August 2010

Acknowledgments

In response to a request by the Department of Homeland Security (DHS), the National Research Council formed an ad hoc committee to assess the current state of the art in private–public sector collaboration dedicated to strengthening community disaster resilience, to identify gaps in knowledge and practice, and to recommend research areas that could be targeted for research investment by the Human Factors Division of the Department of Homeland Security. The committee's charge included organizing a two-day workshop to explore relevant issues and inform the study committee's final recommendations. The workshop was held September 9-10, 2009, in Arlington, Virginia, and engaged a group of approximately 60 participants representing, from different regions of the country, individuals from the private and public sectors and from the research community. The committee thanks those individuals for their contributions.

This report has been reviewed in draft form by persons chosen for their diverse perspectives and technical expertise in accordance with procedures approved by the National Research Council's Report Review Committee. The purposes of this review are to provide candid and critical comments that will assist the institution in making the published report as sound as possible and to ensure that the report meets institutional standards of objectivity, evidence, and responsiveness to the study charge. The review comments and draft manuscript remain confidential to protect the integrity of the deliberative process. We wish to thank the following for their participation in the review of this report:

Ann Patton, Ann Patton Company, LLC, Tulsa, Oklahoma
Carl Maida, University of California at Los Angeles
Daniel Fagbuyi, The George Washington University, Washington, DC
Peter C. Hitt, U.S. Trust Bank of America Private Wealth Management, Baltimore, Maryland
Robert Kates, Independent Scholar, Trenton, Maine

Ron Carlee, International City/County Management Association, Washington, DC
Claudia Albano, City of Oakland, California

Although the reviewers listed above have provided many constructive comments and suggestions, they were not asked to endorse, nor did they see, the final draft of the report before its release. The review of this report was overseen by Ellis Stanley, Dewberry, LLC. Appointed by the Division on Earth and Life Studies, he was responsible for making certain that an independent examination of this report was carried out in accordance with institutional procedures and that all review comments were carefully considered. Responsibility for the final content of this report rests entirely with the authoring committee and the National Research Council.

Contents

Summary

Natural disasters—including hurricanes, earthquakes, volcanic eruptions, and floods—caused over 220,000 deaths worldwide in the first half of 2010 and wreaked havoc on homes, buildings, and the environment. To withstand and recover from natural and human-caused disasters, it is essential that citizens and communities work together to anticipate threats, limit their effects, and rapidly restore functionality after a crisis.

Increasing evidence indicates that collaboration between the private and public sectors could improve the ability of a community to prepare for, respond to, and recover from disasters. Several previous National Research Council reports have identified specific examples of the private and public sectors working cooperatively to reduce the effects of a disaster by implementing building codes, retrofitting buildings, improving community education, or issuing extreme-weather warnings. State and federal governments have acknowledged the importance of collaboration between private and public organizations to develop planning for disaster preparedness and response. Despite growing ad hoc experience across the country, there is currently no comprehensive framework to guide private–public collaboration focused on disaster preparedness, response, and recovery.

To address these concerns, the Department of Homeland Security (DHS) Human Factors Behavioral Sciences Division asked the National Research Council to form a committee of experts to assess the current state of private–public sector collaboration dedicated to strengthening community resilience, to identify gaps in knowledge and practice, and to recommend research that could be targeted for investment (see Box S.1). The committee comprised researchers and practitioners who had expertise in emergency management, local-government management and administration, community collaboration, critical-infrastructure protection, disaster management, and on-the-ground experience establishing and maintaining community resilience initiatives and private–public partnerships. The committee received useful input from practitioners and researchers during a national workshop it convened in September 2009, and published a first report that summarized the major

BOX S.1
Statement of Task

A National Research Council committee will assess the current state of the art in private–public sector partnerships dedicated to strengthening community resilience, identify gaps in knowledge and practice, and recommend research areas that could be targeted for research investment by the DHS Human Factors Division.

In its report, the committee will:

- Identify the components of a framework for private–public sector partnerships dedicated to strengthening community resilience;
- Develop a set of guidelines for private sector engagement in the development of a framework for enhancing community resilience; and
- Examine options and successful models of existing collaborations ranging from centralized to decentralized approaches, and make recommendations for a structure that could further the goal of collaboration between the private and public sectors for the objective of enhancing community resilience.

The study will be organized around a public workshop that explores issues including the following through invited presentations and facilitated discussions among invited participants:

- Current efforts at the regional, state and community levels to develop private–public partnerships for the purpose of developing and enhancing community preparedness and resilience;
- Motivators, inhibitors, advantages and liabilities for private sector engagement in private–public sector cooperation in planning, resource allocation and preparedness for natural and man-made hazards;
- Distinctions in perceptions or motivations between large national-level corporations and the small business community that might influence the formation of private–public sector partnerships, particularly in smaller or rural communities;
- Gaps in current knowledge and practice in private–public sector partnerships that inhibit the ability to develop collaboration across sectors;
- Research areas that could bridge these gaps; and
- Design, development and implementation of collaborative endeavors for the purpose of strengthening the resilience of communities to natural and man-made hazards.

workshop themes. The present report includes the committee's conclusions and guidelines in response to its charge. A key finding of the report is that local-level private–public collaboration is essential to the development of community resilience. Sustainable and effective resilience-focused private–public collaboration is dependent on several basic principles that increase communication among all sectors of the community, incorporate flexibility into collaborative networks, and encourage regular reassessment of collaborative missions, goals, and practices.

DISASTERS

As populations continue to grow and migrate to urban areas, devastation caused by disasters will increase. In developing countries, disasters tend toward a higher rate of fatalities, in part due to inadequate infrastructure, lack of building codes, and poor land use. In the developed world, the cascading consequences of disasters increase as supply chains and critical infrastructure become more interdependent in a global economy. Combined decadal economic and insured losses to natural disasters have increased by a factor of nearly 7 since the 1980s.

As global climate changes, natural disasters, such as hurricanes, coastal storms, floods, droughts, and wildfires, may become more frequent and more intense. Given projections related to climate change, combined with demographic and economic trends that suggest population growth in higher risk coastal areas, the nation could face a future of more disasters, resulting in greater loss of life, greater economic impacts, and greater social disruption. Even in a moderate climate, disasters and technologic disruptions can trigger serious and cascading effects; for example, the 2010 winter snowstorms on the mid-Atlantic coast closed the federal government for five days at an estimated cost of $100 million a day.

The increasing pace of social change, innovation, and technologic advances can combine to create additional vulnerabilities. Regional and global dependencies may make it difficult for individual business operations or entire industries to tolerate disruptions that occur on the other side of the globe. Current inventory and delivery strategies and outsourcing models can result in profitable business, but they leave businesses vulnerable to technology failure. This was the case following the Icelandic volcano eruption in 2010 that grounded a large percentage of global air travel. Local and international commerce worldwide dependent on rapid inventory shipments were severely stressed. For example, commercial flower growers in Africa could not deliver their products to their European markets.

Nationwide, emergency-management policies and systems highlight an all-hazards approach to disaster preparation. Such approaches call for formulated emergency-management responses to likely threats, such as release of hazardous materials, earthquakes, or terrorist attacks with weapons of mass destruction. The committee recognizes the challenges in mobilizing communities against low-probability but high-consequence events, and that particular types of hazards—such as pandemic influenza, bioterrorism, and chemical hazards—require specialized expertise and the development of specialized collaborative subnetworks; however, it also finds that communities prepared for the most common disruptions are those most likely to adapt in the face of more severe or unexpected threats.

COMMUNITY AND COMMUNITY RESILIENCE

Communities are dynamic and respond to changes in population, political leadership, the economy, and environmental factors. Resilient communities can withstand hazards, continue to operate under stress, adapt to adversity, and recover functionality after a crisis. However, community resilience is not just about disasters. The term *resilience* describes the continued ability of a person, group, or system to function during and after any sort of stress. A healthy community with a strong economy, commitment to social justice, and strong environmental standards will be able to bounce back better after a disaster; such communities exhibit a greater degree of resilience. Building and maintaining disaster resilience depends on the ability of a community to monitor change and then modify plans and activities appropriately to accommodate the observed change. The committee finds that private–public collaboration is crucial to the building of networks and trust vital to creating and sustaining healthy, resilient communities.

In considering disaster resilience, a community cannot be defined solely by jurisdictional boundaries because disasters do not fall neatly within geographic limits. In this report, *community* is defined as a group of people who have a common domain of interest—in this case, disaster resilience. The committee finds it very important to engage representatives of the full fabric of the community in decisions related to the full disaster cycle: disaster mitigation, preparedness, response, and recovery. Effective private–public collaboration includes government emergency-response agencies, other public-sector organizations, and all elements of the private sector. The committee defines the private sector to include businesses, nongovernment organizations, volunteer, academic and technical institutions, faith-based organizations, and other civic-minded organizations. Successful collaboration is ideally informed by people from all walks of life, including minorities, the disenfranchised, those with disabilities, children, the elderly, and other populations that are potentially vulnerable. It is essential to have representation for those who deal continually with crises such as poverty, crime, violence, serious illness, and unemployment—the most vulnerable in the community—because survival often takes precedence over issues associated with disaster preparedness and resilience among those members of the community. Engaging the full community in resilience-focused activities, rather than merely providing resources to those who require assistance, allows communities to leverage fully the resources and capacities resident in the community. Through collaboration, participants and those they represent become empowered community members.

THE NECESSITY OF PRIVATE–PUBLIC COLLABORATION

Collaborative arrangements emerge when key public- and private-sector actors recognize that individual and community goals cannot be effectively achieved through independent efforts alone. The private and public sectors each have resources, capabilities, and access to different parts of the community. Through their collective efforts to identify interdependencies, needs, and resources in advance, a community can significantly improve its disaster resilience.

Private–public collaboration for disaster resilience can benefit the entire community, and in ways beyond its disaster-related focus. Collaborative relationships will be more productive and sustainable if they provide incentives, value, and rewards to all stakeholders. In commercial enterprise, for example, profit is important, and the return on investment in resilience-focused private–public collaboration may not be immediately obvious to a business owner. Disaster-related private–public collaboration may benefit business by building trusted networks, providing greater knowledge of interdependencies and local critical infrastructure, and improving coordination with other community stakeholders before, during, and after a disaster. Companies that actively lead such efforts may enjoy greater acknowledgement and standing in the community. Other benefits include communitywide identification of potential hazards, enabling more accurate risk and benefit analyses, and minimizing the consequences of disruption. In addition, by strengthening the resilience of individual businesses, the entire community benefits from a more sustainable economy.

However, without the shared expectation within a community that resilience-focused private–public collaboration is beneficial for the entire community, community resilience will not be easily created or sustained.

A FRAMEWORK FOR RESILIENCE-FOCUSED PRIVATE–PUBLIC SECTOR COLLABORATION

The committee developed a conceptual model for private–public collaboration on the premise that 1) disaster resilience correlates strongly with community resilience; 2) private–public collaboration is based on relationships in which two or more private and public entities coordinate resources toward common objectives; 3) effective collaboration depends on a community-engagement approach; and 4) principles of comprehensive emergency management ideally guide resilience-focused collaboration. The conceptual model, illustrated in Figure S.1, was developed based in large part on community-coalition action theory used in public health applications.

The committee finds that collaboration is best developed in stages and assessed as community networks are developed. Private–public collaboration is more sustainable if it begins as a bottom-up enterprise at the grassroots level—instigated by a leader or organization in the community—rather than dictated top down from a command-and-control structure. The collaborative partnership will ideally reflect and accommodate the unique factors of the community it serves. Such factors include jurisdictional challenges, politics, public policy, geography, local priorities, and access to resources.

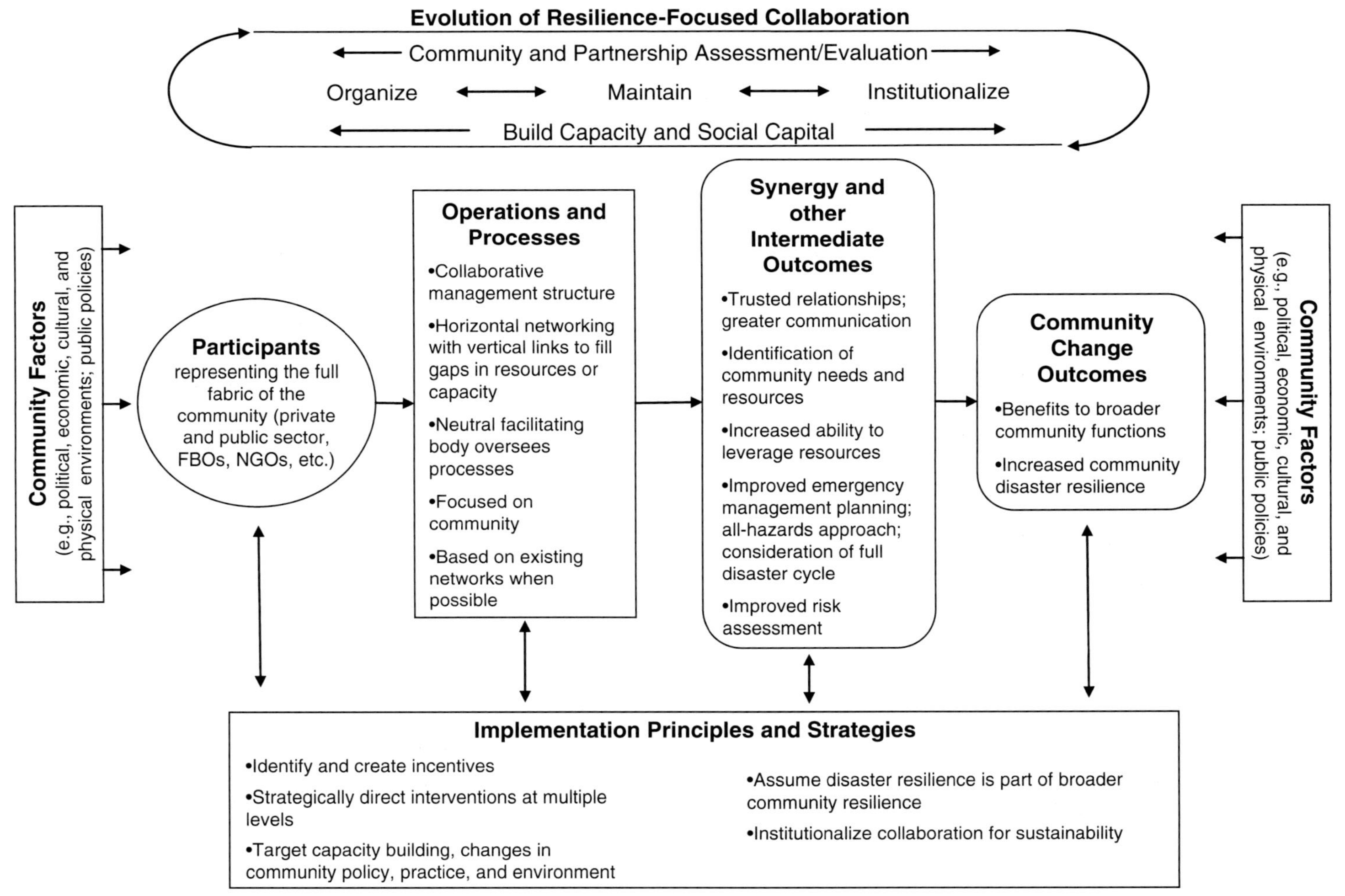

FIGURE S.1 Conceptual Model for Private–Public Sector Collaboration for Building Community Resilience.

Collaboration may begin through the inspiration of one or more community leaders in any sector. Successful growth of a collaborative partnership is most likely if the mission and structure of partnership are developed initially by a core team of community leaders and then broadened to include other key community stakeholders, as capacity and funding are available to ensure stability and effectiveness. Because priorities will be determined by active participants, identifying the right community representatives is a strategic decision. Failure to identify key stakeholders effectively may result in failure to develop the community's full capacity. Inadequate planning with all segments of the population in New Orleans, for example, contributed to the failure to evacuate large portions of the population before Hurricane Katrina. Community-level networking may expand to include existing social networks when feasible. New networks may be needed to reach the disenfranchised or to create greater efficiencies. Networking with higher levels of government or industry—for example, at the state and national levels—is an important means of gaining additional support, but the committee concludes that collaboration is most effective when its leadership is at the local level.

As the collaborative network grows, implementation principles and strategies based on collaborative goals and missions are best decided on collectively to win community acceptance and build trust. Strategies are most successful when they are based on available resources and capacities. It is in a community's interest to design interventions and strategies that can be applied to multiple purposes or are scalable to situations of different proportions; it is a waste of community resources to reinvent the wheel for each new scenario. Resilience-building interventions will be most successful if directed to the entire community and communicated in ways that are meaningful to different populations within the community.

Collaborative goals that effect real change in community policies, practice, and environment are vital, but it is essential that goals also include the sustainability and effectiveness of the collaborative mechanism itself. Sustainable private–public collaboration depends on trust, communication, strong bonds between the private and public sectors, and acceptable returns on investment for all involved. Collaboration requires structure, leadership, and institutional acceptance of the overall mission. The most appropriate structural organization and leadership is representative of community characteristics and common goals. Effective decision making is grounded in trusted relationships and common purpose. Because different community sectors and populations are motivated by different factors, the collaborative structure itself will be strongest if it is trusted and perceived as neutral, nonpartisan, and focused on the greater good of the community. There are examples of successful centralized and decentralized approaches to private–public collaboration, but the committee considers decentralized approaches more conducive to relevant and sustainable resilience-focused collaboration. Regardless of the structure chosen, however, successful collaborative entities often employ staff to serve in a neutral body whose primary function is to facilitate collaboration, activities, and fundraising in advance of a disaster. The experience of these staff ultimately reduces jurisdictional confusion and wrangling after a disaster and allows more efficient pooling of resources and faster recovery.

Synergy in the community will be the result of effective resilience-focused private–public collaboration even before the ultimate goal of increased community disaster resilience is reached. Effective collaboration will increase communication and trust in the community, identify community needs and resources, increase the ability to leverage resources for the benefit of the community, and improve emergency and community planning.

OVERARCHING GUIDELINES

The committee developed a series of guidelines on the basis of its framework and conceptual model intended for those who wish to create an environment supportive of community-level collaboration. The committee was tasked with developing a set of guidelines for private-sector engagement, but finds that the overarching guidelines may be applied by and to all sectors. Effective and sustainable collaboration fosters rather than controls the building of community disaster resilience. It is important to design disaster resilience partnerships themselves to function well in the event of partial or catastrophic failure of community infrastructure. The committee's overarching guidelines are summarized in Box S.2. Challenges to collaboration, however, are inevitable. Successful collaboration is sensitive to the challenges associated with capacity building and access for vulnerable populations; public perception of risk and uncertainty; the difference in scales of organizational operation and scales of needed action; the diverging interests of community stakeholders; trust and information sharing; the need to span organizational boundaries; fragmentation and lack of coordination; and the lack of metrics to measure resilience, the strength of collaboration, and collaboration outcomes.

Though this report addresses primarily community-level private–public collaboration for enhancing disaster resilience, the guidelines are applicable to collaboration—or those wishing to support collaboration—at any level.

RESEARCH DIRECTIONS

Research in many disciplines can be applied to community-level resilience-focused private–public collaboration. However, because most resilience-focused collaborative efforts are largely in nascent stages throughout the nation and because social environments and vulnerability to hazards evolve rapidly, a program of research run parallel to the development of collaborative efforts is imperative, and embedding research within collaborative efforts is

BOX S.2
Overarching Guidelines for Successful Resilience-Focused Private–Public Collaboration

These guidelines can be used in concert with the committee's conceptual model for resilience-focused private–public sector collaboration (Figure S.1) that shows the relationship between collaborative elements and outcomes.

1. Pursue community-level private–public sector collaboration as a fundamental component of community resilience in general and disaster resilience in particular. Resilience-focused private–public collaboration ideally will:

   a. Integrate with broader capacity-building efforts within the community and include all community actors.
   b. Emphasize principles of comprehensive emergency management allowing preparation for all hazards and all phases of the disaster cycle to drive goals and activities.
   c. Function as a system of horizontal networks at the community level, coordinating with higher government and organizational levels.
   d. Develop flexible, evolving entities and establish processes to set goals, conduct continuing self-assessment, meet new challenges, and ensure sustainability.
   e. Institutionalize as a neutral, nonpartisan entity with dedicated staff.

2. Build capacity through communication and training programs for those engaged in private–public collaboration and for the broader community. Resilience-focused private–public collaboration ideally will:

   a. Incorporate capacity building into collaboration from the onset.
   b. Target educational campaigns toward crisis mitigation with goals of community readiness, continuity planning, trust building, risk reduction, and shortened recovery time.
   c. Encourage all organizations in the private and public sectors to commit to organizational resilience through business-continuity measures.
   d. Partner with educational institutions in developing educational campaigns and disseminating information.
   e. Institutionalize the practice of embedding research into resilience-focused private–public sector collaboration by building research directly into existing and future collaborative efforts.

3. Respect well-informed, locally determined all-hazards preparedness and resilience priorities.

4. Develop funding and resource allocation strategies that are flexible in administration.

ideal. The latter would allow the collection of information that could inform collaborative decision making in real time while informing future collaborative efforts.

Series of research and demonstration projects across the nation could be conceptualized as living laboratories, providing opportunities for both researchers and practitioners, and could be designed and undertaken with the explicit goal of documenting effectiveness, costs, and benefits—and the metrics for these variables—and to provide longitudinal and comparative data for future efforts. Below is a set of research initiatives that could be targeted for investment by the DHS and others interested in deepening knowledge on resilience-focused private–public sector collaboration.

- Investigate factors most likely to motivate businesses of all sizes to collaborate with the public sector to build disaster resilience in different types of communities (for example, rural and urban).
- Focus research on how to motivate and integrate community-based, faith-based, and other nongovernment organizations—including those not crisis oriented—into resilience-focused collaboration.
- Focus research on how the emergency-management and homeland security sectors can be moved toward a "culture of collaboration" that engages the full fabric of the community in enhancing resilience.
- Focus research on ways to build capacity for resilience-focused private–public sector collaboration.
- Focus on research and demonstration projects that quantify risk and outcome metrics, enhance disaster resilience at the community level, and document best practices.
- Focus on research and related activities that produce comparable nationwide data on both vulnerability and resilience.
- Establish a national repository and clearinghouse, administered by a neutral entity, to archive and disseminate information on community resilience-focused private–public sector collaboration models, operational frameworks, community disaster-resilience case studies, evidence-based best practices, and resilience-related data and research findings. Relevant stakeholders in all sectors and at all levels should convene to determine how to structure and fund this entity.

A nation is resilient when it is made up of resilient communities. Private–public collaboration is a key step for building such resilience.

CHAPTER ONE

Introduction

The single greatest strength that we possess is the indomitable spirit and capability of the American people. So building a resilient nation doesn't come from a top-down, government-only, command-and-control approach; it comes from a bottom-up approach; it comes from Americans connecting, collaborating; it comes from asking questions and finding new solutions. And it comes from all of us as a shared responsibility.

—Janet Napolitano, Secretary of Homeland Security,
to American Red Cross, July 29, 2009

Secretary of Homeland Security Janet Napolitano and many business executives, leaders of nongovernment organizations (NGOs), and academics conclude: effective private–public collaborations are essential for building community-level disaster resilience. This prompts a series of questions:

- What is resilience?
- To what threats should our communities and our nation be resilient?
- What is the state of resilience-building collaborations across the nation?
- What makes existing partnerships effective?
- By what criteria are partnerships judged, and what is the current state of the art?
- What are the challenges in achieving successful community-level collaboration for disaster resilience?
- What remedies are available?
- What are the essential elements of a framework for effective collaboration?

STATEMENT OF TASK

To date, the private and public sectors have lacked a comprehensive framework to guide their efforts as they collaborate for the purpose of enhancing community disaster resilience. Under the sponsorship of the Department of Homeland Security (DHS), the National Research Council convened a panel of experts to assess the state of the art of private–public sector collaboration dedicated to strengthening community resilience, to identify gaps in knowledge and practice, and to recommend research to be targeted for investment by the DHS Human Factors Behavioral Sciences Division. The committee comprised researchers and practitioners who had expertise in emergency management, local-government management and administration, community and multistakeholder collaboration, critical-infrastructure protection, disaster management, and on-the-ground experience in establishing and maintaining community-resilience initiatives and public–private partnerships. Appendix A presents brief biographies of the committee members. The committee's statement of task, as provided by the DHS, is shown in Box 1.1. The committee received useful input during a national workshop that it convened on September 9–10, 2009, and prepared a summary of the major themes discussed in the workshop (NRC, 2010a).

Collaboration between the private and public sectors could improve the ability of a community to mitigate, prepare for, respond to, and recover from natural or human-caused disasters. Past reports from the National Research Council have identified innovative, collaborative organizational structures that could enhance the diverse community interests in matters of national concern (e.g., NRC, 1998, 2006). Others have identified specific efforts where the private and public sectors have worked cooperatively on measures that reduce the effects of disaster—such as implementing building codes, retrofitting buildings, and issuing extreme-weather warnings—and identified candidates for such collaboration, such as risk-based insurance premiums and model land-use practices (e.g., Mason, 2006; Jones Kershaw, 2005). Recognizing that a community's ability to respond to and recover from disaster depends partly on the strength and effectiveness of its social networks, DHS sponsored a 2009 National Research Council workshop on how social network analysis—the study of complex human systems—can reveal the structure of existing networks so that a community can design or improve its networks for the purpose of building community resilience (Magsino, 2009).

To help the reader understand the concepts deliberated by the committee, this chapter provides working definitions for key terms such as "resilience" and "community." Examples of disasters that challenge community resilience are provided, beginning with a brief discussion of the financial burden associated with disasters. The committee then briefly examines disaster management policy in the United States and the role of private–public collaboration in building community resilience. A description of the committee's approach to addressing its charge and a description of the report organization completes this chapter.

BOX 1.1
Statement of Task

A National Research Council committee will assess the current state of the art in private–public sector partnerships dedicated to strengthening community resilience, identify gaps in knowledge and practice, and recommend research areas that could be targeted for research investment by the DHS Human Factors Division.

In its report, the committee will:

- Identify the components of a framework for private–public sector partnerships dedicated to strengthening community resilience;
- Develop a set of guidelines for private sector engagement in the development of a framework for enhancing community resilience; and
- Examine options and successful models of existing collaborations ranging from centralized to decentralized approaches, and make recommendations for a structure that could further the goal of collaboration between the private and public sectors for the objective of enhancing community resilience.

The study will be organized around a public workshop that explores issues including the following through invited presentations and facilitated discussions among invited participants:

- Current efforts at the regional, state and community levels to develop private–public partnerships for the purpose of developing and enhancing community preparedness and resilience;
- Motivators, inhibitors, advantages and liabilities for private sector engagement in private–public sector cooperation in planning, resource allocation and preparedness for natural and man-made hazards;
- Distinctions in perceptions or motivations between large national-level corporations and the small business community that might influence the formation of private–public sector partnerships, particularly in smaller or rural communities;
- Gaps in current knowledge and practice in private–public sector partnerships that inhibit the ability to develop collaboration across sectors;
- Research areas that could bridge these gaps; and
- Design, development and implementation of collaborative endeavors for the purpose of strengthening the resilience of communities to natural and man-made hazards.

WHAT IS RESILIENCE?

The term *resilience* is encountered in many disciplines, but no definition is common to all. Different elements or attributes of resilience are emphasized, but all definitions speak in a general way to the continued ability of a person, group, or system to adapt to stress—such as any sort of disturbance—so that it may continue to function, or quickly recover its ability to function, during and after stress.

The committee charge included focus on "community resilience." In its work, the committee relied on a definition of *resilience* put forward by Norris and others (2008), who describe it as the ability of groups, such as communities and cities, to withstand hazards or to recover from such disruptions as natural disasters. Building and maintaining resilience depend on the ability of a group to monitor changes and to modify its plans to deal with adversity appropriately. Similarly, John Plodinec has observed that the ability of a community to recover after a disaster is greater if resilience was implicitly or explicitly considered by members of the community as an inherent and dynamic part of the community (CARRI, 2009). He understands that a resilient community is one that anticipates threats, mitigates potential harm when possible, and prepares to adapt in adversity. Such communities more rapidly recover and restore functionality after a crisis. He has also indicated that a community's ability to compare itself to other communities with respect the ability to adapt to adversity is important because it can help identify needed improvements.[1]

Community resilience thus refers generally to the continued ability of a community to function during and after stress. Implicit in discussion of building community disaster resilience in this report is that all sectors of a community (government, private for-profit, private nonprofit, and citizens) can and should participate in building resilience through all phases of disaster: mitigation, preparedness, response, and recovery.

COMMUNITY AS MORE THAN JURISDICTION

The term *community* is defined differently by different people when they consider disaster preparedness, response and recovery planning, and implementation. Defining communities by geographic boundaries ignores the reality that disasters do not respect jurisdictions. Community-level collaboration intended to address disruptions must draw on the full array of diverse social networks in which residents and public and private entities are engaged. These are not defined exclusively by, or confined to, jurisdictional boundaries. Definitions of *community* based on jurisdictional boundaries may lead to a static idea of what constitutes a community; in reality, communities are dynamic and ever-changing. Similarly, while a community may extend beyond geographical and political boundaries, it might also be defined as something much smaller. In large municipalities—such as Los Angeles, California, or New York City, New York—individuals may be tied to a sense of community that is much smaller and of more immediate scale.

Etienne Wenger defines a community as "a group of people for whom the domain of interest is relevant" (Wenger, 1998). The committee expands Wenger's "group of people"

[1]J. Plodinec, Community and Regional Resilience Institute, personal communication, June 28, 2010.

to include the full fabric of a community and all its partners. The "domain of interest" in this report is community disaster resilience. Seeing communities as dynamic and connected with entities beyond jurisdictional boundaries does not negate the importance of collaboration that reflects the needs, priorities, and economies of the geographic communities and regions the collaborative networks serve.

The phrase "full fabric of the community" is used throughout this report and is integral to the committee's definition of community, particularly in the context of disasters and the role of collaboration at the local level. Community disaster mitigation, planning, response, and recovery require the active involvement of local government, but the attention and engagement of federal, state, regional, and tribal governments are also essential, as are private-sector energies and assets (Edwards, 2009). The committee defines the private sector broadly and comprehensively as including large and small for-profit corporations and also nongovernment, volunteer, academic, faith-based, and other entities that help define the social life and stability of a community. The committee understands that private–public collaboration to achieve community disaster resilience hinges on the notion that disruptions such as disasters tear at all or portions of a community's social fabric.

TO WHAT MUST WE BE RESILIENT?

A myriad of potential disasters puts communities at risk. Natural and human-caused disasters result in public health emergencies suffering, loss of life, damage to economies, and damage to community environments. Individuals and institutions often fail to perceive that hazards may pose unacceptable risk to their communities and ways of life. Further, individuals and institutions often fail to accept their role in reducing that risk. The next sections describe some types of disasters that could affect communities. These hazards include natural disasters, public health emergencies, human-caused disasters, disasters caused by cyber vulnerabilities or by emerging technological and business practices, and climate change. Some of these risks may be greater for some communities than others, and communities may face other hazards not discussed in this report, including those related to the very real effects of economic recessions and unemployment.

Losses from disaster can devastate communities and nations. Natural and human-caused disasters claimed 240,000 lives in 2008 and nearly 15,000 lives in 2009[2] worldwide and led to economic losses of approximately US$268 billion and US$62 billion, respectively (see Figure 1.1). Swiss Reinsurance Company estimated in early 2010 that the cost of natural disasters alone in 2010 could reach US$110 billion worldwide (Swiss Re, 2010).

[2]Nearly 9,000 people died or were missing because of natural disasters in 2009; the others were victims of human-caused disasters, i.e., major events associated with human activities (excluding war, civil war, and warlike events).

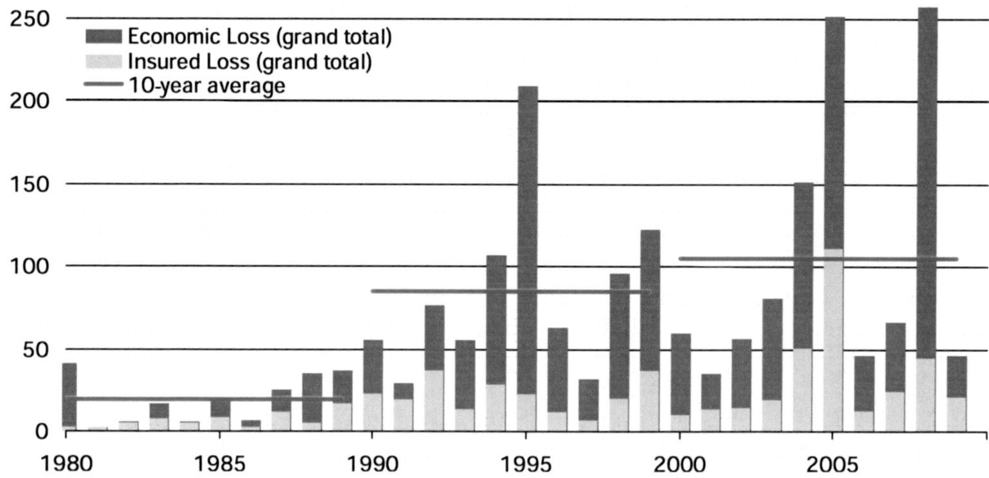

FIGURE 1.1 Natural-catastrophe losses worldwide, 1980–2009, in billions of U.S. dollars (indexed to 2009). The spike corresponding with 1995 reflects largely the Kobe earthquake. The 2005 spike represents the effects of Hurricane Katrina. The 2008 spike correlates with the earthquake in China and Hurricane Ike in the United States. SOURCE: Swiss Re, sigma catastrophe database.

This was determined after the earthquakes in Haiti and Chile, but before, for example, the massive flooding that occurred in many parts of Asia. Figure 1.1 shows a steady rise in the financial losses associated with natural disasters worldwide from 1980 to 2009. By comparison, Table 1.1 lists the human and economic losses to major human-caused disasters worldwide in 2009, according to loss category.

Many research and policy communities acknowledge the threat of disasters and associated economic losses and have sought to reduce socioeconomic vulnerability to, for example, climate and weather-related hazards. They include groups interested in disaster-risk reduction, climate-change adaptation, environmental management, and poverty reduction. The work of those groups, however, has been fragmented, and the groups have worked largely independently of one another, so they have had only small success in reducing vulnerability (Thomalla et al., 2006). In later sections of this report, the committee will make the case for an "all hazards" approach to building community resilience, which means understanding all hazards that pose a threat to community but focusing attention on the ones most likely to occur. It is an underlying assumption of the committee that a resilient community prepared for one kind of disaster will be able to adapt when faced with another.

TABLE 1.1 List of Major Losses Worldwide in 2009 According to Loss Category

	Number of Events	Percent of Total	Dead or Missing	Percent of Total	Insured Losses[a] (USD)	Percent of Total
Natural disasters	133	46.2%	8,977	60.2%	22,355	85.1%
Human-caused disasters	155	53.8%	5,939	39.8%	3,915	14.9%
Major fires and explosions	30	10.4%	756	5.1%	1,605	6.1%
Aviation and space disasters	15	5.2%	783	5.2%	752	2.9%
Maritime disasters	39	13.5%	2,146	14.5%	1,359	5.2%
Rail disasters (incl. cableways)	10	3.5%	70	0.5%	1	0.0%
Mining accidents	11	3.8%	544	3.6%	43	0.2%
Collapse of buildings/bridges	10	3.5%	410	2.7%	86	0.3%
Miscellaneous[b]	40	13.9%	1,230	8.2%	69	0.2%
Total	288	100.0%	14,916	100.0%	26,270	100.0%

[a] Property and business interruption, excluding liability and life insurance losses
[b] Includes social unrest, terrorism, and "other miscellaneous losses"
SOURCE: Swiss Re (2010).

Natural Disasters

According to the Federal Emergency Management Agency (FEMA), there have been 66 declarations of disaster in the United States in 2010 (as of September); in contrast, there were 59 disaster declarations in all of 2009.[3] Insured natural-disaster losses in the United States exceeded $11 billion in 2009 (Munich Re, 2009). In the decade 2000–2009, natural disasters in the United States caused over $350 billion in economic losses, or an average of $35 billion per year (Munich Re, 2009). For many in harm's way, financial losses can be catastrophic—the loss of home or savings for retirement. Distribution of declared U.S. disasters in the last decade[4] indicates that most Americans will be affected by disaster sometime in their lives. The loss is equivalent to $1,200 for every American over the 10-year period. Combined decadal economic and insured losses to natural disasters have increased by a factor of nearly 6 since the 1980s, as illustrated in Figure 1.2. By contrast, the U.S. Gross Domestic Product has only doubled during this same period.[5]

[3]See www.fema.gov/news/disaster_totals_annual.fema (accessed May 17, 2010).
[4]See www.gismaps.fema.gov/recent.pdf (accessed September 7, 2010) for a map of Presidential Disaster Declarations.
[5]See www.data360.org/dataset.aspx?Data_Set_Id=354 (accessed September 7, 2010).

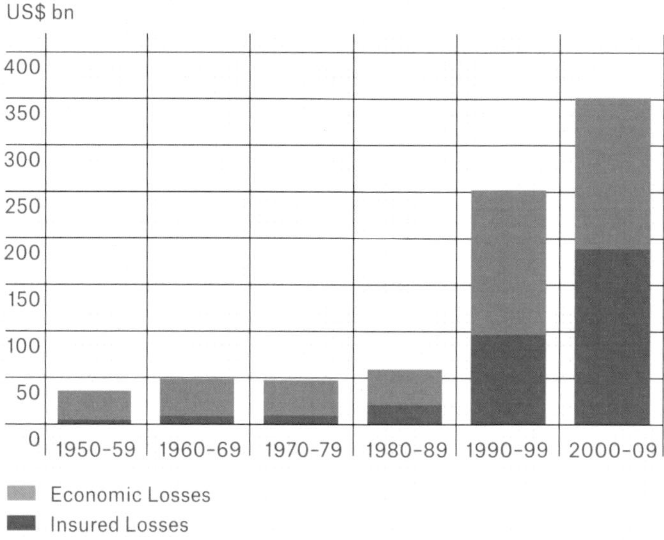

FIGURE 1.2 Estimated economic and insured losses to natural disasters (in 2009 dollars) in the United States per decade. SOURCE: ©2010 Münchener Rückversicherungs-Gesellschaft, Geo Risks Research, NatCatSERVICE. Munich Re (2009).

Earthquakes, hurricanes, floods, droughts, and other weather-related events; landslides; and volcanic hazards can affect communities well beyond those physically affected by the event. This is due, in part, to increased interconnectedness between local and national communities. Human and economic losses associated with these impacts are steadily increasing, in part because of increasing population densities. The 10 costliest disasters since 1950 occurred in the years 1992–2010 (Wirtz, 2010). Figure 1.1 indicates that losses to natural catastrophes worldwide have risen substantially, from an average of about US$20 billion in the 1980s to an average of over US$100 billion in the 2000s (Swiss Re, 2010). The global death toll from moderate earthquakes in the coming decades is predicted to average 8,000–10,000 per year. Individual catastrophic earthquakes are predicted to cause the decadal average to exceed 50,000 per year (Bilham, 2009). The 2004 Indian Ocean earthquake and subsequent tsunami, which resulted in over 220,000 deaths,[6] reminds us that the Pacific United States is vulnerable to similar events.

[6]See earthquake.usgs.gov/earthquakes/eqinthenews/ (accessed September 10, 2010).

Public Health Emergencies

Communities are vulnerable to public health emergencies that may arise from natural or human causes. These include emergencies associated with pandemics, bioterrorism, mass casualties caused by terrorist or accidental incidents, chemical emergencies, emergencies arising from natural disasters and severe weather, radiation emergencies, and threats to water and food security including water- and food-borne diseases. Community vulnerability to a pandemic was brought to immediate attention in 2009 following a worldwide outbreak of the potentially deadly influenza A (H1N1) virus. On June 11, 2009, the World Health Organization (WHO) declared that a global pandemic of the H1N1 virus was underway, and by June 19, all 50 states, the District of Columbia, Puerto Rico, and the U.S. Virgin Islands reported cases of the virus. On August 10, 2010, the WHO declared an end to the outbreak.[7] The total global or national cost of the 2009 H1N1 pandemic has not been calculated with certainty, and the effect of the pandemic was less severe than some predicted. However, some research indicates the average annual cost of influenza in the United States is approximately $10.4 billion in direct medical costs, with a total economic burden of $87.1 billion (Molinari et al., 2007).[8]

The H1N1 virus reminded the nation how vulnerable communities are to public health disasters. Given the increase in travel among U.S. residents, even small communities are not immune to the dangers of a pandemic. Increasing population in urban centers means a greater risk of spread of disease. Part of an all-hazards approach to community resilience is consideration of all manner of threats to public health that can affect the health, economy, and proper functioning of the community.

Human-Caused Disasters

The nation's communities are also vulnerable to disasters caused by failures of technology and by willful acts of terrorism. Disasters resulting from the development of energy resources and the disposal of their wastes have been a fact of life for many communities since the industrial revolution. In modern times, the failure of a coal-waste impoundment dam in West Virginia after heavy rains resulted in 125 deaths and an estimated $50 million in property damage in what has become known as the Buffalo Creek flood of 1972 (NRC, 2002). Several other coal-waste impoundment failures have occurred since 1972, including a 2008 failure in Kingston, Tennessee, that released over a billion gallons of coal-waste slurry onto communities and into watersheds. The latter was described as the most serious toxic disaster of its kind to have occurred in the United States (Dewan, 2008).

[7]See www.cdc.gov/h1n1flu/background.htm (accessed September 13, 2010).

[8]Lost productivity from missed work days and lost lives comprise the bulk of the economic burden of influenza.

Toxic disasters can also result from the energy extraction and transport industries. The 1989 Exxon Valdez oil spill is ranked among history's most devastating marine accidents (NRC, 2003), having affected over 1,100 miles of coastline, wildlife, and communities. The social and environmental effects of that spill are still apparent over 20 years later. In April 2010, an oil-rig explosion in the Gulf of Mexico resulted in the death of 11 workers and released tens of thousands of gallons of oil a day into the Gulf for three months, amounting to the largest oil spill ever in U.S. waters (McCoy and Salerno, 2010). The long-term environmental, health, and economic effects of this disaster have yet to be determined, but the Gulf Coast of the United States is already feeling the economic burden; a preliminary analysis by the Dun & Bradstreet Corporation found that the spill may impact 7.3 million active businesses throughout Alabama, Florida, Louisiana, Mississippi, and Texas, affecting 34.4 million employees and $5.2 trillion in sales volume (D&B, 2010). Although the flow of oil was stopped in late July, the National Oceanic and Atmospheric Administration (NOAA) kept a large portion of the Gulf of Mexico closed to commercial and recreational fishing for the remainder of the summer. Figure 1.3 illustrates areas of the Gulf that were closed from June until September 2010.

Nuclear energy production and waste disposal also pose risks. The nuclear reactor meltdown of the Chernobyl nuclear power station in Ukraine in 1986 caused the evacuation and resettlement of 336,000 people from the area (UNSCEAR, 2000). The number of those struck with illness related to radiation is not known, but it is estimated that about 4,000 of the 600,000 people most highly exposed will suffer fatal radiation-induced cases of cancer. Another 5,000 cases of cancer in peripheral populations will probably also be diagnosed (Mettler, 2006). No one is permitted to live within 17 miles of the reactor (Bell, 2006).

Acts of violence and terrorism affect our nation and its communities. The terrorist attacks of September 11, 2001, which caused nearly 3,000 deaths (The 9/11 Commission, 2004), are among the deadliest disasters ever to occur on U.S. soil and have resulted in numerous societal changes in communities and nations around the world. The interdependence of different types of critical infrastructure was made obvious. For example, after the attack in New York City, water-main breaks flooded rail tunnels, a commuter station, and a facility that housed all the cables for one of the world's largest telecommunication nodes. Trading on the New York Stock Exchange was halted for 6 days because of failure of communication infrastructure (O'Rourke, 2007).

Cyber Failure and Cyber Attacks

Cyber infrastructure refers to infrastructure based on integrated distributed computer, information, and communication technology; it includes not only the electronic systems themselves—composed of the hardware and software that process, store, and communicate data—but also on the information contained in these systems (NSF, 2003; DHS, 2009).

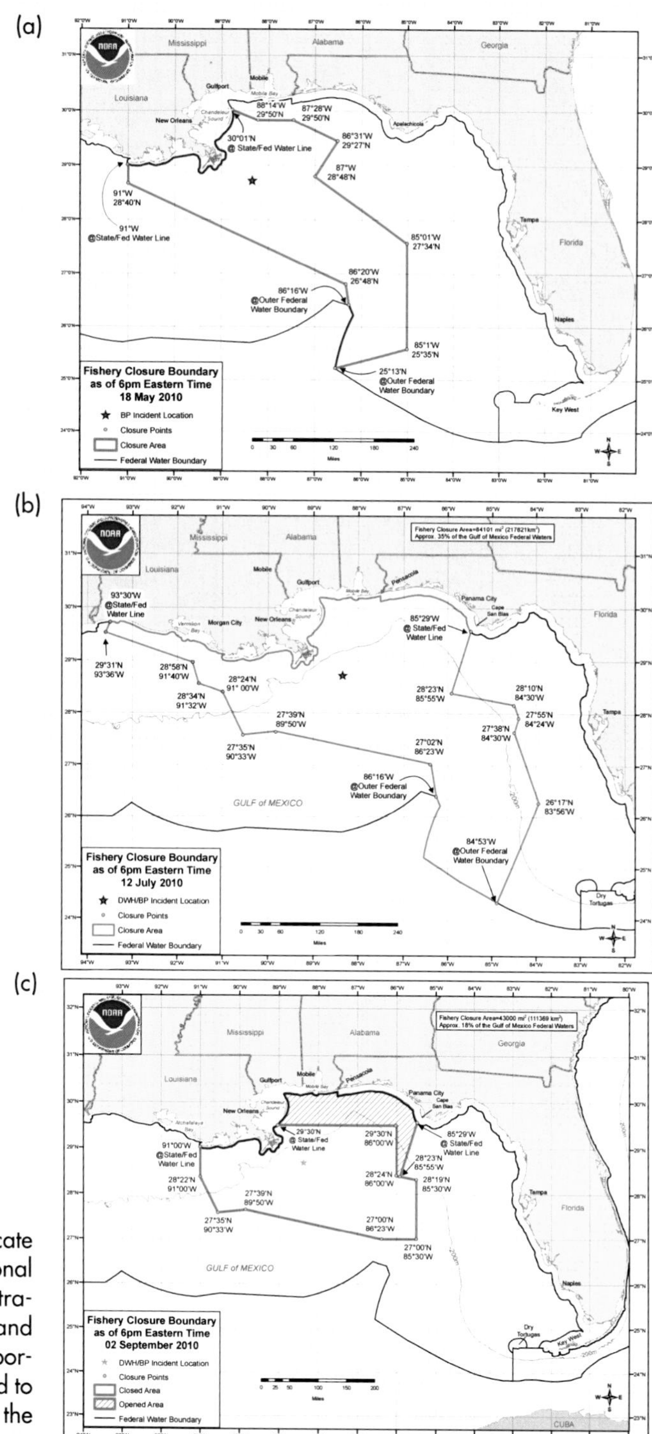

FIGURE 1.3 Red boundaries indicate areas closed to fishing by the National Oceanic and Atmospheric Administration on (a) May 18, (b) July 12, and (c) September 2, 2010. The shaded portion in (c) indicates the area reopened to fishing. The star on each map locates the leaking well. SOURCE: NOAA.

The U.S. economy and national security depend heavily on the global cyber infrastructure. Military, police, firefighters, and other emergency services providers rely on computers, information networks, and the Global Positioning System (GPS) to carry out missions and respond to crises. GPS, for example, is an important timing reference for the national power grid and for telecommunications, including telephone systems, the Internet, and cell phones in this country. As cyber dependency and interconnectedness increase, so does the potential for cyber failure to spread quickly and have debilitating impacts on local and state communities (DHS, 2009).

Cyber failure can result from either natural events or malicious cyber attacks. For example, electromagnetic pulses from solar storms can have disastrous consequences, as happened in 1989 when a severe magnetic storm overloaded the power grid in Quebec, causing millions of dollars in damage and leaving millions of people without power (OCIPEP, 2003). According to an estimate by the Metatech Corporation, a long-term, wide-area blackout caused by an extreme space weather event could cost as much as $2 trillion during the first year, with full recovery requiring 4 to 10 years (NRC, 2009).

Attacks on cyber infrastructure by government, criminal, or terrorist groups or individuals are also a growing concern. A report issued in 2001 for the U.S. Department of Transportation described risks to information and communications infrastructure vulnerabilities affecting civilian aviation, maritime, and surface transportation as a result of loss or degradation of GPS signal (Volpe Center, 2001). That report—almost a decade old—also described risks to transportation cyber infrastructure, many of which are relevant today. A survey in June 2009 found that cybercrime in the previous two years cost Americans more than $8 billion (Consumer Reports, 2009). Popular news outlets have reported that inexpensive GPS jammers that fit in a shirt pocket are available online for purchase (though illegal in the United States), and can be used to disrupt GPS reception and confuse emergency responders who rely on GPS for communication and logistical operations (e.g., Brandon, 2010).

Risk of cyber failure, whether due to natural but largely predictable atmospheric disruptions, mechanical failure, failure of software operation, or malicious intent, is a serious and growing issue for communities across the country.

Climate Change

A National Research Council study indicates that the global climate is changing and that temperatures have risen nearly 2°F (1°C) in the last 50 years (NRC, 2010b). Expert projections of trends related to climate change and variation, as cited in recent reports from the Intergovernmental Panel on Climate Change (e.g., IPCC, 2007a), together with demographic and economic trends that suggest population growth in higher-risk coastal areas show that the United States could face a future of increased loss of life, economic costs, and social disruptions from disasters. Water quality could be drastically affected in communities around the world (IPCC, 2008) with substantial consequences for individuals, businesses, communities, and nations if complex and tightly coupled social and infrastructure systems are affected. Even a moderate climate event, natural disaster, or technologic disruption can trigger serious cascading effects. The 2010 winter snowstorms along the mid-Atlantic coast, for example, closed the federal government for 5 days, with an estimated cost of about $100 million per day (MacAskill, 2010).

Some weather events and extremes, such as hurricanes, coastal storms, floods, droughts, and events that they cause or exacerbate (such as wildfires) may become more frequent, widespread, or intense during the 21st century as a result of climate change (e.g., NRC, 2010b; IPCC, 2007a). As extreme events become more intense or frequent or occur in different locations, their economic and social costs will increase (IPCC, 2007b). Communities around the country will need to anticipate vulnerability to climate change and adopt adaptation strategies to reduce that vulnerability (NRC, 2010b).

Projected increases in population and changes in migration patterns may alter the composition of many communities. The populations of several large American cities, for example, are expected to increase. In the United States, the South and the West are the most heavily populated and fastest-growing regions, and that growth is expected to continue in coming decades (Beach, 2002). Some 53 percent of the American population already lives within 50 miles of a coast (Markham, 2008).

Business Practices and Technologic Evolution

The increasing pace of social change, economic innovation, and technologic advance combine to create potential unanticipated vulnerability. Therefore, disasters of the past are of limited value as a guide to the future. For example, the ways in which we conduct business and efficiencies developed during the last decades—such as outsourcing and "just-in-time" inventory and delivery strategies—result in more profitable business models, but may leave organizations vulnerable. Such efficiencies reduce not only waste but profit margin. Regional and global interdependence may make it difficult for individual business operations or entire industries to tolerate disruptions associated with disasters that take place even in different parts of the world. The ash eruptions of Eyjafjallojökull Volcano in Iceland, for example, affected air traffic and therefore commerce around Europe and the world in April and May 2010 (USGS, 2010). Local businesses in communities that count on immediate air shipments of inventory were stressed: commercial growth of African flowers for European markets is a well-publicized example, and larger cascading adverse economic and social effects may result (ITC, 2010).

DISASTER-MANAGEMENT POLICY

This section provides background and a brief overview of emergency-management policy in the United States to give context for the findings and conclusions presented in the report. The committee makes no recommendations with respect to emergency-management policy. The committee briefly describes the importance of the private sector to disaster management; gives a brief history of disaster management policy—particularly as it relates to approaches to hazards and the role of community-level disaster-related private-public partnerships; and describes the role of local communities in emergency management and the relationship between local and federal emergency managers.

In the United States, the private and public sectors both play a role in disaster management and are integral to the governing policy framework. The private sector supplies many services—such as water, power, communication networks, transportation, medical care, and security—before, during, and after a disaster. The health of the U.S. economy depends on large and small businesses and, in turn, their roles in globalization and rapid technologic advances (Bonvillian, 2004). Critical infrastructure is owned and managed largely by private entities, but existing private–public collaboration related to managing risk and building resilience could be strengthened, and collaboration could be encouraged in communities where there is little or none.

Elements of U.S. disaster-management policy are reflected in legislation and initiatives, including the Stafford Act;[9] the Disaster Mitigation Act of 2000;[10] the Post-Katrina Emergency Management Reform Act;[11] such presidential directives as Homeland Security Presidential Directives 5 and 8;[12] and past and current federal disaster plans and initiatives, such as the Federal Response Plan,[13] the National Response Plan,[14] and the National Response Framework (FEMA, 2008). The legislation and plans reinforce the all-hazards comprehensive emergency-management approach (e.g., considering the full disaster cycle) that has been in effect for three decades. Current presidential directives, policy documents, the National Preparedness Guidelines,[15] the National Response Framework, the National Recovery Framework,[16] the National Incident Management System,[17] and operational and implementing documents also reflect that longstanding practice.

[9]See www.fema.gov/about/stafact.shtm (accessed June 20, 2010).

[10]See www.fema.gov/library/viewRecord.do?id=1935 (accessed June 20, 2010).

[11]See www.dhs.gov/xabout/structure/gc_1169243598416.shtm (accessed June 20, 2010).

[12]See www.dhs.gov/xabout/laws/gc_1214592333605.shtm and www.dhs.gov/xabout/laws/gc_1215444247124.shtm (accessed June 20, 2010).

[13]See biotech.law.lsu.edu/blaw/FEMA/frpfull.pdf (accessed June 20, 2010).

[14]See www.scd.hawaii.gov/documents/nrp.pdf (accessed June 20, 2010).

[15]See www.fema.gov/library/viewRecord.do?id=3773 (accessed June 20, 2010).

[16]See www.fema.gov/recoveryframework/ (accessed June 24, 2010).

[17]See www.fema.gov/emergency/nims/ (accessed June 20, 2010).

The history of emergency management reveals an evolution of institutions and their roles. In the 19th century, disasters and response were viewed as the purview of private charities. In the middle of the 20th century, emergency management focused on nuclear war, civil defense, and increasing government involvement in comprehensive emergency-management functions (Rubin, 2007; FEMA, 2005). During the 1990s, FEMA's innovative but no longer active Project Impact program recognized the vital role of the private sector in all aspects of disaster mitigation (see Box 1.2). FEMA provided funds to more than 250 communities for mitigation and preparedness activities while promoting local autonomy in how the funds were used to reduce risk.[18] The Disaster Mitigation Act of 2000 provided communities with more incentives for predisaster mitigation through the federally funded risk-reduction program. By July 2008, over 17,000 local jurisdictions had mitigation plans that used an active community-engagement approach and that were implemented with federal guidance (Topping, 2009). Amendments to the Stafford Act after Hurricane Katrina in 2005 direct funds toward the mitigation of future federally declared disasters (CRS, 2006). DHS was established in response to the September 11, 2001, attacks.[19] The primary mission of the agency when it was established was to reduce vulnerability and prevent terrorist attacks, but DHS is also responsible for reducing disaster vulnerability more generally.

Emergency management in the United States is based on an approach in which communities are encouraged to mitigate, prepare for, respond to, and recover from disasters at the local level. The encouragement has had limited success as local jurisdictions struggle to accomplish their responsibilities overseeing the daily operations of their communities. In reality, towns, cities and counties often rely on the capacity of the federal government to act as a first responding partner when crises evolve beyond a conventional emergency. The disparity of expectations among varied levels of government can create operational gaps when extreme events occur.

Federal government support for local and state-level activities ranges from limited seed funding for risk reduction and preparedness to larger amounts for specific antiterrorist measures. Emergency-management policies and systems highlight the importance of all-hazards planning,[20] which calls for formulated responses to specific types of events, such as a release of hazardous materials, an earthquake, or a terrorist attack with weapons of mass destruction. That comprehensive principle has guided activities in the emergency-management community since the late 1970s (Whittaker, 1978; NGA, 1979): communities manage risks posed by natural and man-made hazards through preparation, response, and recovery. For a time after the terrorist attacks of September 11, 2001, the management of events arising from acts of terrorism took priority over other concerns (FEMA, 2005;

[18]See www.fema.gov/news/newsrelease.fema?id=8895 (accessed June 20, 2010).

[19]See www.dhs.gov/xabout/laws/law_regulation_rule_0011.shtm (accessed June 20, 2010).

[20]See www.fema.gov/txt/help/fr02-4321.txt (accessed Feb. 26, 2010).

BOX 1.2
Project Impact

In 1997, Congress first appropriated funds for the direct purposes of funding mitigation activities for disasters.[a] With this appropriation, the Federal Emergency Management Agency (FEMA) created a pilot program called *Project Impact: Building Disaster Resistant Communities.* Project Impact placed emphasis on and dedicated resources to the community level and led efforts to mitigate hazards. Community-level decision making was promoted. Communities were required to secure the commitment of local governments, nongovernmental organizations (NGO), and businesses. An educational component to raise awareness was also required. FEMA provided funding to form public and private partnerships within the individual communities. Project Impact envisioned four steps to building a disaster-resistant community:

(1) Building partnerships by organizing a disaster-resistant community planning committee including business and industry, public works and utilities, volunteer and community groups, government, and education, health care, and workforce representatives.
(2) Assessing a community's risks and vulnerabilities.
(3) Identifying mitigation priorities, measures, and resources and taking action
(4) Communicating progress and maintaining collaborative involvement and support for long-term initiatives.

One Project Impact example of success is Tulsa, Oklahoma. Through community effort, Tulsa instituted long-term mitigation activities to reduce flood frequency and severity. Efforts included improving and maintaining channels and detention storage basins and clearing more than 1,000 buildings from floodplains.[b] Despite the termination of Project Impact in 2002, private–public sector collaboration to improve community resilience continues today through an NGO called Tulsa Partners, Inc.[c]

Project Impact was initiated in 1997 with a $2 million appropriation. The program received $30 million in 1998 and $25 million during 1999–2002. In each successive year, new communities were selected in each of the 50 states so that by 2000 Project Impact communities numbered over 250. Most communities received seed-funding grants from FEMA. In February 2001, Congress approved the Bush Administration's proposal to eliminate Project Impact, less than five years after its inception. The administration sought to create a program to carry out mitigation efforts directly.

---

[a] McCarthy F.X. and N. Keegan. 2009. FEMA's Pre-Disaster Mitigation Program: Overview and Issues. Washington, DC: Congressional Research Service. July 10. 25 pp.

[b] See www.emergencymgmt.com/disaster/Project-Impact-Initiative-to.html (accessed August 31, 2010).

[c] See www.tulsapartners.org/About (accessed August 31, 2010).

Haddow and Bullock, 2005)—a trend that was reversed to some degree after the 2005 hurricane season and the devastating effects of Hurricane Katrina.[21]

Disaster management, however, is often confined to first-responder experts, whose specific expertise alone often cannot address conditions that are the consequence of extreme events like natural or human-caused disasters. Nor may they be knowledgeable on appropriate response to nonphysical events such as economic recession and unemployment that can have devastating effects on a community. While U.S. disaster management concepts are intended to be comprehensive, their application is not. Policies that guide federal programs are a helpful foundation for what could be comprehensive risk management at the national level. State and local government would benefit from collaborative relationships with their federal counterparts to yield practical results in the field at the local level.

Planning and training activities for disaster management underpin the system at all levels of government, but full implementation is difficult to achieve. We prepare to respond with the inherent assumption that if we are prepared to respond quickly, efficiently, and effectively, recovery will follow naturally (CARRI, 2009). Expanding on that concept, Harvard Kennedy School research has explored disaster-management practice and has suggested strategic improvements in social welfare with more balanced investments in advance recovery planning and risk reduction (Leonard, 2010). This approach, when applied, promises improved outcomes in disaster situations because it pairs response activities with risk reduction. Eventually, resilience results from conditions that foster nimble and responsive actions in advance of disruptive events (Leonard and Howitt, 2010). The comprehensive risk-management approach provides the nation with a commonly understood and effective system of incident response to and early recovery from most disasters. The system can be flexible and adaptable, as demonstrated by specific problems identified and addressed during major disasters. Improvement plans are ideally made by governments in response to shortcomings identified during particular disasters. Application and implementation of policies is not always effective, as evidenced by poor response when disaster occurs. Emergency and disaster managers and responders may apply "just-in-time" practices to situations that warrant more complex and adaptive action.

COLLABORATION FOR RESILIENCE

Collaboration occurs through a variety of formal and informal arrangements. The committee's use of the term *collaboration* refers to cooperative action. In this report, unless otherwise specified, the terms *partnership*, *coalition*, *network*, *joint venture*, and *alliance* refer to various types of organizations or mechanisms that enable collaboration in the broadest sense, regardless of the formality of the arrangement. Different sectors may identify these

[21] See www.dhs.gov/xfoia/archives/gc_1157649340100.shtm (accessed June 21, 2010).

terms differently. In the private sector, for example, partnerships and joint ventures imply contractual arrangements between organizations that include business plans with formalized marketing, finance, and operations components. The terms may be applied differently in other sectors.

How do notions of collaboration play into the building of disaster resilience? The term *collaborate* is defined by Merriam-Webster as "to work jointly with others or together especially in an intellectual endeavor" and "to cooperate with an agency or instrumentality with which one is not immediately connected."[22] Human affairs (and their history) can be understood in terms of collaboration. The human condition and the prospects for humanity's future are determined not just by demographics, geography, the growth and nature of economies, the advance of science and technology, or a conjunction of critical moments in history with the emergence of heroic individuals. They depend in at least equal measure on how people, institutions, and sectors of institutions engage and work with each other in the array of human concerns and aspirations on scales from the local to the regional to the national to the global. The concept of collaboration is an organizing principle or lens with which to view society and suggests how things can be accomplished (Wright, 2001).

Unavoidable and sometimes unpredictable extreme natural events may result in disasters because of the decisions people make regarding societal land use and development, public safety and health, economic growth, protection of the environment, and geopolitical stability (e.g., Mileti, 1999). With proper decision making and preparation, however, disasters can be avoided or their effects mitigated. Events can be anticipated, and resilient societies factor them into planning and action. Both researchers and practitioners increasingly appreciate the intersection of collaboration and disasters and are paying greater attention to private–public collaboration to build community disaster resilience (CARRI, 2009).

THE COMMITTEE'S APPROACH TO ITS TASK

The recent popular work *A Paradise Built in Hell: The Extraordinary Communities that Arise in Disaster* (Solnit, 2009) describes how disaster can be the crucible for a community's transformation. This committee's task was to identify how communities can encourage that transformation, correct resource deficiencies, adopt beneficial public policies, and exercise practical means to elicit functional, community-based partnerships well before disaster strikes. To determine how this could be done, the committee convened, as part of its charge, a national workshop that brought together researchers and others from for-profit organizations, various levels of government, and citizen and volunteer organizations actively involved in collaborative approaches to community disaster resilience (NRC, 2010a). The

[22]See www.merriam-webster.com/dictionary/collaboration (accessed May 25, 2010).

committee held three additional meetings to obtain input and to deliberate. Appendix B includes agendas for all open sessions.

The goals of the workshop were to elicit the best thinking on private–public sector partnership as a means to enhance community resilience at the levels most affected by natural and human-caused disasters. Discussions in the workshop played an essential part in the committee's development of its recommendations. Having diverse stakeholders participate in discussion helped to bring to light a number of verifiable best practices as applied in numerous successful partnerships that crossed sector and jurisdictional boundaries.

The committee developed some presumptive principles that became the organizing themes for its workshop:

- Collaboration is essential for community disaster resilience.
- Private–public sector collaboration should include interjurisdictional organizations, diverse industry sectors, nongovernment organizations, and all elements of the community—not just government and the for-profit sector.
- Community disaster resilience is essential for all phases of predisaster and post-disaster planning and action—from mitigation through long-term recovery.

Committee members wanted to test—through workshop discussions and testimony from practitioners, community leaders, and subject-matter experts—whether those themes were common to experience and practice. The workshop affirmed the principles, as documented in the workshop report (NRC, 2010a). Communities, academic institutions, and professional experts from across the country held similar notions about resilience and about how it can be promoted at the regional and local levels.

This study is one of multiple activities related to disaster resilience undertaken by the National Research Council (e.g., National Earthquake Resilience: Research, Implementation, and Outreach;[23] NRC, 2006, 2007, 2009; Magsino, 2009; McCoy and Salerno, 2010). The Institute of Medicine (IOM) of the National Academies has addressed the issue of public health emergencies and community resilience. In a letter report addressing research priorities in emergency preparedness and response for public health systems, the IOM has called for research related to (1) the design and implementation of training for improved public health preparedness; (2) improved communications for the effective exchange of information with diverse audiences; (3) sustainable preparedness and response systems to identify factors that affect a community's successful response to a crisis with public health consequences; and (4) criteria and metrics for the measurement of effectiveness and ef-

[23] A current study. See www8.nationalacademies.org/cp/projectview.aspx?key=49048 for more information (accessed May 31, 2010).

ficiency in the evaluation of public health emergency preparedness, response, and recovery (IOM, 2006). That committee found that:

> The organization and operations of effective systems of public health preparedness need to be constituted to cope with a wide range of threats—the all-hazards approach—including catastrophic health events. . . . include state, local, tribal, and federal public health agencies; practitioners from emergency response and health-care delivery systems; communities, homeland security and public safety, health-care delivery systems, employers and business, the media, academia, and individual citizens. . . . Public health emergencies will vary in scale, timing, predictability, and the potential to overwhelm routine capabilities and to disrupt the provision of daily life and health-care services. (IOM, 2006: 13)

This study is the first, however, to focus solely on community-level resilience and especially on the role of private–public collaboration in enhancing community-level disaster resilience.

REPORT ORGANIZATION

This report provides the reader with both a conceptual framework for community-level, resilience-focused private–public collaboration and guidance on how such collaboration may be established. The theoretical basis for private–public collaboration is provided in Chapter 2. The chapter lays out the committee's primary assumptions and justifications regarding resilience and collaboration, the committee's framework, and finally its conceptual model outlining the major elements of resilience-focused private–public collaboration. It also addresses how collaboration can work at the local or community level but in a multilevel context that spans local, state, and national organizations in both the private and public sectors. Chapter 3 provides guidelines to develop, implement, and evaluate collaboration at all levels. Chapter 4 summarizes challenges to the formation and maintenance of private–public collaboration including those associated with increasing capacity and access of vulnerable populations; perceptions of risk and uncertainty; scales of collaboration; trust and information sharing; diverging interests; lack of coordination; and lack of outcome measures. Chapter 5 identifies research that could advance knowledge and understanding the committee considers crucial to inform strategies for forming, maintaining, and supporting private–public collaboration.

REFERENCES

Beach, D. 2002. *Coastal Sprawl: The Effects of Urban Design on Aquatic Ecosystems in the United States.* The Pew Charitable Trusts. April 8. Available at www.pewtrusts.org/our_work_report_detail.aspx?id=30037 (accessed June 20, 2010).

Bell, R. 2006. *Disasters: Wasted Lives, Valuable Lessons.* Wyomissing, PA: Tapestry Press.

Bilham, R. 2009. The Seismic Future of Cities. Twelfth Annual Mallet-Milne Lecture. July 17. *Bulletin of Earthquake Engineering.* DOI 10.1007/s10518-009-9147-0. Available at cires.colorado.edu/~bilham/MalletMilneXIIBilham.pdf (accessed June 20, 2010).

Bonvillian, W. 2004. Meeting the New Challenge to U.S. Economic Competitiveness. *Issues in Science and Technology.* Available at www.issues.org/21.1/bonvillian.html (accessed June 20, 2010).

Brandon, J. 2010. GPS Jammers Illegal, Dangerous, and Very Easy to Buy. March 17. Available at www.foxnews.com/scitech/2010/03/17/gps-jammers-easily-accessible-potentially-dangerous-risk/ (accessed September 14, 2010).

CARRI (Community and Regional Resilience Institute). 2009. Toward a Common Framework for Community Resilience. Draft in progress. Presented to the Community and Resilience Roundtable, Washington, DC, December 1.

Consumer Reports. 2009. "Boom time for cybercrime." *Consumer Reports Magazine.* June. Yonkers, NY: Consumers Union of U.S., Inc. Available at www.consumerreports.org/cro/magazine-archive/june-2009/electronics-computers/state-of-the-net/overview/state-of-the-net-ov.htm (accessed September 10, 2010).

CRS (Congressional Research Service). 2006. Federal Emergency Management Policy Changes after Hurricane Katrina: A Summary of Statutory Provisions. November 15. Washington, DC: Congressional Research Service. Available at www.fas.org/sgp/crs/homesec/RL33729.pdf (accessed June 20, 2010).

D&B (The Dun & Bradstreet Corporation). 2010. 2010 Deepwater Horizon Oil Spill: Preliminary Business Impact Analysis for Coastal Areas in the Gulf States. June 7. Available at www.dnbgov.com/pdf/DNB_Gulf_Coast_Oil_Spill_Impact_Analysis.pdf (accessed September 7, 2010).

DHS (Department of Homeland Security). 2009. National Infrastructure Protection Plan: Partnering to Enhance Protection and Resiliency. Washington, DC: U.S. Department of Homeland Security. Available at www.dhs.gov/xlibrary/assets/NIPP_Plan.pdf (accessed August 5, 2010).

Dewan, S. 2008. Tennessee Ash Flood Larger than Initial Estimate. *The New York Times.* December 26. Available at www.nytimes.com/2008/12/27/us/27sludge.html (accessed Feb. 26, 2010).

Edwards, W. 2009. Engaging the full-fabric of communities. *CARRI Blog.* Oak Ridge, TN: Community & Regional Resilience Institute. Available at blog.resilientus.mediapulse.com/2009/07/09/engaging-the-full-fabric-of-communities/ (accessed June 24, 2010).

FEMA (Federal Emergency Management Agency). 2005. "Chapter 1 – Introduction to Crisis, Disaster, and Risk Management Concepts." *Emergency and Risk Management Case Studies Textbook.* Emmitsburg, MD: Emergency Management Institute. Available at training.fema.gov/EMIWeb/edu/emoutline.asp (accessed June 20, 2010).

FEMA (Federal Emergency Management Agency). 2008. National Response Framework. Washington, DC: U.S. Department of Homeland Security. Available at www.fema.gov/pdf/emergency/nrf/nrf-core.pdf (accessed March 11, 2010).

Haddow, G. and J. Bullock. 2005. The Future of Emergency Management. June. Washington, DC: Institute for Crisis, Disaster and Risk Management, George Washington University. Available at www.training.fema.gov/emiweb/edu/docs/emfuture/Future%20of%20EM%20-%20The%20Future%20of%20EM%20-%20Haddow%20and%20Bullock.doc

IOM (Institute of Medicine). 2006. Modeling community containment for pandemic influenza: A letter report. Washington, DC: The National Academies Press. Available at http://www.nap.edu/catalog.php?record_id=11800 (accessed September 10, 2010).

IPCC (Intergovernmental Panel on Climate Change). 2007a. *Climate Change 2007: Synthesis Report.* Geneva, Switzerland: Intergovernmental Panel on Climate Change. Available at www.ipcc.ch/publications_and_data/publications_ipcc_fourth_assessment_report_synthesis_report.htm (accessed June 20, 2010).

IPCC (Intergovernmental Panel on Climate Change). 2007b. *Climate Change 2007: The Physical Science Basis—Summary for Policymakers and Technical Summary.* Cambridge, UK: University Press. Available at www.ipcc.ch/publications_and_data/ar4/wg1/en/contents.html (accessed June 20, 2010).

IPCC (Intergovernmental Panel on Climate Change). 2008. *Climate Change and Water.* Geneva, Switzerland: Intergovernmental Panel on Climate Change. Available at www.ipcc.ch/meetings/session28/doc13.pdf (accessed June 30, 2010).

ITC (International Trade Centre). 2010. The Impact of European Airspace Closures on African Horticultural Exports. Geneva, Switzerland. Available at www.intracen.org/docman/PRSR15431.pdf (accessed July 19, 2010).

Jones Kershaw, P., ed. 2005. Creating a Disaster Resilient America: Grand Challenges in Science and Technology: Summary of a Workshop. Washington, DC: The National Academies Press.

Leonard, H. B. 2010. Creating a Better Architecture for Global Risk Management: A Proposal to the World Economic Forum Global Redesign Initiative. Global Agenda Council on Catastrophic Risks, World Economic Forum. Available at www.hks.harvard.edu/var/ezp_site/storage/fckeditor/file/pdfs/centers-programs/programs/crisis-leadership/GAC%20Catastrophic%20Risks%202009%20White%20Paper.pdf (accessed September 5, 2010).

Leonard, H. B. and A. M. Howitt. 2010. Chapter 2: Acting in Time Against Disaster. In *Learning from Catastrophes: Strategies for Reaction and Response*. Eds. H. Kunreuther and M. Useem. Upper Saddle River, NJ: Wharton Press School Publishing, 2010.

MacAskill, E. 2010. Washington DC paralysed by snow for fifth working day in a row. Guardian.co.uk. Available at www.guardian.co.uk/world/2010/feb/11/washington-snow (accessed June 30, 2010).

Magsino, S. 2009. *Applications of Social Network Analysis for Building Community Disaster Resilience: Workshop Summary*. Washington, DC: The National Academies Press.

Markham, V. D. 2008. U.S Population, Energy & Climate Change. New Canaan, CT: Center for Environment and Population. Available at www.cepnet.org/documents/USPopulationEnergyandClimateChangeReportCEP.pdf (accessed February 26, 2010).

Mason, B., ed. 2006. Community Disaster Resilience: A Summary of the March 20, 2006 Workshop of the Disasters Roundtable. Washington, DC: The National Academies Press.

McCoy, M. A., and J. A. Salerno. 2010. *Assessing the Effects of the Gulf of Mexico Oil Spill on Human Health: A Summary of the June 2010 Workshop*. Washington, DC: The National Academies Press.

Mettler, F. A. 2006. Chernobyl's Living Legacy. *IAEA Bulletin* 47(2). Available at www.iaea.org/Publications/Magazines/Bulletin/Bull472/pdfs/chernobyl.pdf (accessed June 8, 2010).

Mileti, D., 1999. *Disasters by Design: A Reassessment of Natural Hazards in the United States*. Washington, DC: The Joseph Henry Press.

Molinari, N.-A. M., I. R. Ortega-Sanchez, M. L. Messonnier, W. W. Thompson, P. M. Wortley, E. Weintraub, and C. B. Bridges. 2007. The annual impact of seasonal influenza in the US: Measuring disease burden and costs. *Vaccine* 25(27): 5086-5096.

Munich Re. 2009. TOPICS GEO Natural catastrophes 2009: Analyses, assessments, positions. U.S. Version. Munich, Germany.

NGA (National Governors Association). 1979. Comprehensive Emergency Management: A Governor's Guide. Washington, DC: NGA. Available at training.fema.gov/EMIWeb/edu/docs/Comprehensive%20EM%20-%20NGA.doc (accessed June 20, 2010).

The 9/11 Commission. 2004. The 9/11 Commission Report: Final Report of the National Commission on Terrorist Attacks Upon the United States. Washington, DC: U.S. Government Printing Office. Available at www.9-11commission.gov/report/911Report.pdf (accessed August 4, 2010).

Norris, F. H., S. P. Stevens, B. Pfefferbaum, K. F. Wyche, and R. L. Pfefferbaum. 2008. Community resilience as a metaphor: Theory, set of capacities, and strategy for disaster readiness. *American Journal of Community Psychology* 41(1-2):127-150.

NRC (National Research Council). 1998. *Toward an Earth Science Enterprise Federation: Results from a Workshop*. Washington, DC: National Academy Press.

NRC (National Research Council). 2002. *Coal Waste Impoundments: Risks, Responses, and Alternatives*. Washington, DC: The National Academies Press.

NRC (National Research Council). 2003. *Oil in the Sea III: Inputs, Fates, and Effects*. Washington, DC: The National Academies Press.

NRC (National Research Council). 2006. *Facing Hazards and Disasters: Understanding Human Dimensions*. Washington, DC: The National Academies Press.

NRC (National Research Council). 2007. *Improving Disaster Management: The Role of IT in Mitigation, Preparedness, Response, and Recovery*. Washington, DC: The National Academies Press.

NRC (National Research Council). 2009. *Severe Space Weather Events—Understanding Societal and Economic Impacts: A Workshop Report*. Washington, DC: The National Academies Press.

NRC (National Research Council). 2010a. *Private–Public Sector Collaboration to Enhance Community Disaster Resilience: A Workshop Report.* Washington, DC: The National Academies Press.

NRC (National Research Council). 2010b. *Adapting to the Impacts of Climate Change.* Washington, DC: The National Academies Press.

NSF (National Science Foundation). 2003. Revolutionizing Science and Engineering Through Cyberinfrastructure: Report of the National Science Foundation Advisory Panel on Cyberinfrastructure. Arlington, VA: NSF. Available at www.nsf.gov/od/oci/reports/toc.jsp (accessed September 10, 2010).

OCIPEP (Office of Critical Infrastructure Protection and Emergency Preparedness). 2003. Threat Analysis: Threats to Canada's Critical Infrastructure. No. TA03-001. March 12. Available at www.publicsafety.gc.ca/prg/em/ccirc/_fl/ta03-001-eng.pdf (accessed September 10, 2010).

O'Rourke, T. D. 2007. Critical Infrastructure, Interdependencies, and Resilience. *The Bridge* . Volume 37(1). Available at www.caenz.com/info/RINZ/downloads/Bridge_Article.pdf (accessed June 20, 2010).

Rubin, C. B., ed. 2007. *Emergency management: The American experience, 1900-2005.* Fairfax, VA: Public Entity Risk Institute.

Solnit, R. 2009. *Paradise Built in Hell: The Extraordinary Communities that Arise in Disaster.* London, UK: Viking Press.

Swiss Re. 2010. Natural catastrophes and man-made disasters in 2009. *Sigma No. 1/2010.* Zurich, Switzerland: Swiss Reinsurance Company Ltd.

Thomalla, F., T. Downing, E. Spanger-Siegfried, G. Han, and J. Rockströom, 2006. Reducing hazard vulnerability: towards a common approach between disaster risk reduction and climate adaptation. *Disasters* 30(1): 39-48.

Topping, K. 2009. Toward a National Disaster Recovery Act of 2009. *Natural Hazards Observer* 33(3): 1-8. Available at www.colorado.edu/hazards/o/archives/2009/jan_observerweb.pdf (accessed June 20, 2010).

UNSCEAR (United Nations Scientific Committee on the Effects of Atomic Radiation). 2000. Annex J: Exposures and effects of the Chernobyl Accident. In *Sources and Effects of Ionizing Radiation*, Vol. II: Effects. Available at www.unscear.org/docs/reports/annexj.pdf (accessed July 20, 2010).

USGS (U.S. Geological Survey). 2010. Eyjafjallojökull, Ash, and Eruption Impacts. Available at volcanoes.usgs.gov/publications/2010/iceland.php (accessed June 24, 2010).

Volpe Center. 2001. Vulnerability Assessment of the Transportation Infrastructure Relying on the Global Positioning System: Final Report. Prepared for the Office of the Assistant Secretary for Transportation Policy, U.S. Department of Transportation. Available at ntl.bts.gov/lib/31000/31300/31379/17_2001_Volpe_GPS_Vulnerability_Study.pdf (accessed September 15, 2010).

Wenger, E. 1998. *Communities of Practice: Learning, Meaning, and Identity.* Cambridge, UK: Cambridge University Press.

Whittaker, H. 1978. State comprehensive emergency management: final report of the Emergency Preparedness Project. Washington, DC: Center for Policy Research, National Governors' Association.

Wirtz, A. 2010. Careful Data Management. *D+C Development+Cooperation* 51(5): 240-242. Available at www.inwent.org/ez/articles/174547/index.en.shtml (accessed June 24, 2010).

Wright, R. 2001. *Nonzero: The Logic of Human Destiny.* London, UK: Vintage Books.

CHAPTER TWO

A Conceptual Framework for Resilience-Focused Private–Public Collaborative Networks

The committee's charge included the development of a framework for private–public collaboration to enhance community disaster resilience. Any single template or checklist would not sufficiently address the full array of needs for collaboration of all communities around the country or the diverse threats they face. The committee therefore sought to develop an overarching conceptual framework that would provide the context in which collaborative efforts are best undertaken. The framework laid out in this chapter focuses on organizational aspects of encouraging and enabling private–public collaboration and on the processes and strategies for institutionalizing effective communitywide collaboration. To create the framework, the committee explored theoretical concepts and models and related literature. The resulting conceptual model is the basis of the specific guidelines and examples provided later in this report.

Three themes are presented in this chapter. The first is the theoretical necessity for private–public collaboration focused on building community resilience. The committee describes the assumptions on which its theoretical framework is based, discusses the role of collaboration in comprehensive emergency management and capacity building, and explains what disaster resilience means for a community. The second theme is the theoretical basis for successful collaboration. The chapter delves into concepts such as creating incentives, planning perspectives, and the advantages of decentralized decision making processes. Levels of engagement are also addressed. The final theme is the committee's conceptual model for resilience-focused private–public collaboration and describes the elements therein. Organizational aspects of the framework will be presented in Chapter 3.

BASIC PRINCIPLES THAT SHAPE THE CONCEPTUAL FRAME

The overarching conceptual frame that guides this report is derived from research in several disaster-related disciplines and from guidance the committee received at its workshop (NRC, 2010). The framework rests on the following assumptions:

- Disaster resilience correlates strongly with community resilience, including economic, environmental, health, and social-justice factors.
- Private–public collaboration is based on collaborative relationships in which two or more private and public entities pool and coordinate the use of complementary resources through the joint pursuit of common objectives.
- Effective collaboration ideally encompasses the full fabric of the community and is representative of all walks of life—including minorities, the impoverished or disenfranchised, children, and the elderly—so a community-engagement approach is essential for the success of resilience-focused collaboration.
- Principles of comprehensive emergency management, incorporating an all-hazards approach, guide resilience-focused collaboration-building efforts.

The framework adopted by the committee assumes that disaster resilience is closely linked with broader capacity-building strategies aimed at long-term community and environmental sustainability. The relationship between disaster resilience and sustainability is directly proportional: communities that suffer high losses in disasters are often the ones that have paid little attention to overall sustainability issues, and communities that actively plan for a more sustainable future are more likely to achieve disaster resilience. Thus, resilience-focused collaboration is likely to be most effective when integrated with and built on broader community functions, including those associated with public health and safety, economic viability, housing quality, infrastructure development, and environmental quality. As multiple workshop participants noted, community resilience involves more than disaster response (NRC, 2010).

Why Collaborate?

Scholarship focusing on the evolution of institutional forms emphasizes that such activities as the production and delivery of goods and services are seldom undertaken by single large corporations or by vast government bureaucracies. Rather, various parties that own or manage different types of resources work in concert to produce and provide goods and services.

The same societal trends influence efforts related to disaster-loss reduction. Taking an example from the homeland security arena, the Department of Homeland Security has a statutory responsibility to protect the critical infrastructure of the United States, but much critical infrastructure is owned and managed by private entities.[1] Protection can be achieved through collaboration among government and private entities. At the city and county levels, various public agencies—such as local emergency-management agencies and police, fire, and emergency medical services agencies—each have specific response-related roles, but they cannot meet their objectives alone. Reliance on broad participation by private entities—such as private hospitals, debris-removal contractors, the Red Cross, the Salvation Army, other nonprofit entities that provide aid to disaster victims, and privately owned utilities—is essential.

Public policy scholars also note that collaborative approaches are invariably needed to address large, complex problems, particularly ones that can be categorized as "wicked problems" (e.g., Rittel and Webber, 1973; Rayner, 2006). Wicked problems have several characteristics: they are extremely complex, people who offer solutions often disagree, it is difficult to address different aspects of these types of problems incrementally because they are tightly interwoven, and they are never solved "once and for all." Analysts note that wicked problems are often intractable because the parties that should provide solutions are often the ones that helped create the problems. The scale and complexity of wicked problems demand collaboration among agencies, organizations, sectors, jurisdictions, and disciplines and fields of expertise. Examples of wicked problems are those associated with climate change, homeland security, and disaster reduction.

Although organizations increasingly rely on collaboration to achieve their goals and tackle wicked problems, collaborators are still independent actors who generally cannot be compelled to work with one another. Instead, potential partners interact, learn about one another, and weigh the costs and benefits of affiliating with other parties before agreeing to work together. Appropriate forms of governance for their collaborative activities can then be developed.

Businesses and other private-sector organizations are the foundation of the U.S. economy. Critical infrastructure providers include those that provide lifeline services such as power, water, and natural gas, as well as those that provide banking and financial services, information technology and telecommunication services, transportation, food and agricultural services, and health care services. Communities in the United States could not function without those services. Success in providing those services—and the success of many private-sector organizations more generally—often depends on the efficiency of the logistics and supply chain management. Large and small businesses and organizations that represent business interests have therefore become critical elements in the community social fabric. Collaboration with nongovernment organizations (NGOs) and private voluntary and faith-based organizations enables government agencies to build capacity. All elements

[1]It is estimated that about 80 percent of critical infrastructure in the United States is in private hands (DHS, 2009; TISP, 2006).

of the private sector are equal partners in successful community resilience-building efforts because of their function in every community.

The need for private–public collaboration relates directly to the evolution of governance in the United States and around the world. Contemporary developed societies are diverse, complex, and to a large extent information-driven, and they are much different from the bureaucracies and hierarchies that characterized the industrial age (Agranoff and McGuire, 2003). They require societal institutions that work compatibly and collaboratively. Outsourcing and contracting by government agencies have become common in the provision of government services; these practices bring together actors from the private and public sectors in complex relationships. Fast-moving economic and technologic processes require businesses to be flexible in forming alliances and joint ventures (Moynihan, 2005). Collaboration is essential for the provision of all types of goods and services and for the common welfare, including community disaster resilience.

Collaborative arrangements emerge because of the recognition that individual and collective goals are more likely to be achieved through collaborative rather than independent efforts. Collaboration is founded on trusted relationships, information sharing, incentives, and common goals, so facilitating and sustaining effective collaboration is challenging in a "command-and-control" environment. Benefits of collaboration are widely documented, and there is a substantial literature on collaborative management (e.g., McGuire, 2006), public administration (e.g., Vigoda, 2002), and collaborative emergency management (e.g., Waugh and Streib, 2006). The committee finds that the principles and approaches developed in such fields are vitally important to shaping resilience-enhancing collaboration, strategies, and goals.

Collaboration for Comprehensive Emergency Management

The committee considered literature on community engagement strategies and processes, including scholarship in such fields as public health and environmental protection. Lessons learned in those and related disciplines have implications for disaster-loss reduction. Under the principles of comprehensive emergency management, collaboration may focus on building community-level resilience to all types of disruptive events, from those most likely to occur to the rare, worst-case scenarios. The committee recognizes that particular types of hazards—such as pandemic influenza, bioterrorism, and chemical hazards—may require specialized capabilities and the development of specialized collaborative networks within networks. But the committee takes the position that communities prepared for the most common disruptions are also more likely to adapt in the face of more unusual threats. At the same time, the committee advocates for specialized planning by those communities with known unusual but identifiable risks—for example, risks associated with proximity to nuclear or chemical facilities.

The committee also concludes that a collaborative framework that addresses challenges across the full hazards cycle—from pre-event mitigation measures through efforts aimed at long-term recovery—is most likely to succeed at building resilience. It recognizes that not every community can take on all stages of disaster management and some may focus on one or two elements of the hazards cycle. It is important, however, to recognize how all the stages of the disaster cycle are linked and to plan accordingly.

Collaboration and Capacity Building

Private–public sector collaboration is an essential component of building capacity in communities. Collaborative relationships often begin with local organizers who have identified specific community needs. The process continues by mobilizing key leaders and relevant stakeholders in the community. Communication strategies and mechanisms that enable information sharing are critical to expanding collaboration to the broader community. Training programs in the use of communication tools may be useful to the organizers, as well as training on how to facilitate communitywide collaboration.

Community in the Context of Disaster Resilience

Effective resilience-focused collaborative networks are representative of the communities they serve, but they can also be coordinated with individuals and organizations outside the community. Ideally, collaboration includes representatives from local, state, and federal agencies; small and large businesses; nonprofit and faith-based organizations; academicians, researchers, and educational institutions; the mass media; civic and neighborhood organizations; technical experts; volunteers; and other diverse community stakeholders. The wealthy and the poor, the politically influential and those who are not, and both majority and minority populations would likewise be engaged. Identifying the critical points of contact for all constituencies in the community makes communication and outreach most effective. Doing so helps identify and mobilize the different perspectives and capabilities needed to address challenges fully and provide resources for building capacity.

Specific resources may not be available in some communities, and this confirms the importance of extending the reach of community beyond jurisdictional or geographic boundaries. When a community needs specific resources, collaborative networks may expand to incorporate regional stakeholders to fill the gaps. Disasters ignore jurisdictional and geographic boundaries, so communities will benefit by looking beyond such boundaries when building community disaster resilience.

Disaster management is a holistic function that cannot be successful if it does not engage the full fabric of the community. William Waugh, an emergency-management expert, testified before a subcommittee of the House of Representatives that the national emergency-management system is made up of the local emergency-management offices, response agencies, and faith-based and other community organizations and that it is essential to engage these networks of private, public, and nonprofit organizations (Waugh, 2007). He also noted that the surge capacity during emergencies is often provided by ad hoc volunteer groups and individual volunteers:

> We have a long history of volunteerism in emergency management in the United States and should always expect that volunteers will be a significant segment of our disaster response operations. Most fire departments today are still volunteer organizations. Most search and rescue is done by neighbors, family members, and friends. Faith-based and secular community groups increasingly have their own disaster relief organizations and the capabilities of those organizations are increasing rapidly. The point is that we have a system in place for dealing with large and small disasters that is heavily reliant upon local resources and local capacities.

The United States has long been a nation in which people and groups mobilize on a voluntary basis to achieve community objectives. Benjamin Franklin, for example, acted on his belief that voluntary cooperative action was good for the community when he established the first volunteer fire departments, public lending libraries, and fire insurance companies in Philadelphia, Pennsylvania (Heffner, 2001). His writing may have influenced others, such as Alexis de Tocqueville, who wrote in the early 1830s of formal and informal associations that provide the context for citizens to participate in their communities. The principles that undergird citizen involvement and collaboration are the same as those that form the foundation of democracy itself (Pickeral, 2005). Collaborative networks are tools for involving the full fabric of the community and, by doing so, make disaster resilience easier to achieve.

Defining Disaster Resilience in Terms of Community Resilience

A community becomes more disaster resilient through a conscious effort to do so. Community disaster resilience is best achieved through broad efforts that address economic, social, and environmental issues; disaster resilience is seldom achieved independent of broader community interests. To optimize community disaster resilience, however, it is essential for community stakeholders to form a common understanding of what community disaster resilience comprises. In this section of the report, the committee describes the relationship between community disaster resilience and community resilience, and how this relationship may be leveraged through private–public collaboration.

There is little empirical evidence to show that communities that incorporate general capacity-building strategies, including those to enhance social capital, into community planning strategies are in a better position to withstand disasters than their counterparts.

Research in areas related to social capital and economics, however, indicates that social capital is vitally important to organization performance (e.g. Burt, 2000). In an organization or group of organizations, networking and social capital control who has access to what information and when that information can be used advantageously. The committee extends this relationship to communities and performance during and following disaster. A well-connected community may be in a better position to share information to be used in crisis situations with its stakeholders. Strategically doing so will likely improve its stakeholders' resilience. Communities recognize the value of collaboration for capacity building for a variety of purposes and are characterized by robust and active engagement between civil-society, government, and private-sector organizations. Participants in the committee's information-gathering workshop stressed this concept. Disaster resilience is a byproduct of more general activities designed to improve the social and economic well-being of community residents. Being prepared for and surviving adversity are prerequisites of a healthy community. Ron Carlee, former manager of Arlington County, Virginia, and currently chief operating officer and director of strategic initiatives at the International City/County Management Association, emphasized during the committee's workshop that "resiliency is not just for disasters . . . we need to build functional communities that provide quality of life everyday" (see Box 2.1).

Disaster-resilient communities, as a normal part of community functioning, prepare and plan to respond to and recover from disasters that are most likely to occur. Response and recovery take into account and benefit from the full fabric of the community, engaging

BOX 2.1
Resilience: Not Just for Disasters

Communities that are factionalized, divisive, and suspicious of public and private institutions as a matter of routine are not likely to become models of collaboration during a disaster. Communities that have the best potential for achieving disaster resilience

- are committed to social equity and inclusion
- are economically and environmentally sustainable
- build a vision that is shared by residents and institutions—public, nonprofit, and private
- have a sense of place
- unite people around values and purpose

SOURCE: R. Carlee, Arlington County, Presentation to the Workshop on Private–Public Sector Collaboration to Enhance Community Disaster Resilience, Sept. 10, 2010.

all elements of the population in efforts to increase resilience. They are led by residents, organizations, and community partners that work collectively to achieve resilience and that identify and connect the networks and systems relevant to the resilience goals. As outlined in the National Response Framework (FEMA, 2008), local communities are ultimately responsible for managing hazards and disasters, and that responsibility requires the engagement of all community stakeholders in the private and public sectors, and faith-based organizations and NGOs (FEMA, 2008). Although leadership and incentives from the state and national levels may help communities to become disaster-resilient, community resilience is more sustainable when it is pursued from the ground up, is locally led and managed, and includes the full fabric of the community.

Mobilizing a Community Toward Resilience

Numerous challenges confront efforts to create disaster-resilient communities. In disadvantaged communities or during perilous economic times, daily survival often takes precedence over planning for low-probability natural disasters. The contrasting impacts of the 2010 earthquakes in Haiti and Chile make vivid the importance of building resilience (see Box 2.2). The scale of devastation in Haiti was far greater than in Chile, in large part because of the level of advance preparation for a known risk.

A community-organization approach may be a means of successfully mobilizing communities toward resilience. Minkler and Wallerstein (1999:30) define community organiz-

BOX 2.2
Earthquakes in Haiti and Chile

The 2010 earthquakes in Haiti and Chile illustrate how disaster preparedness can alter the outcome of similar catastrophic events. The earthquake and resulting tsunamis in Chile, although severe, were not unusual for the region, which has experienced 13 earthquakes of magnitude 7.0 or greater since 1973. The country was relatively well prepared for the event. In contrast, the people of Haiti were largely unaware of earthquake risks—the region last experienced a major earthquake in 1860—and poverty, poor building design and construction, and a lack of building standards led to the huge losses suffered in that country. Both earthquakes affected approximately 1.8 million people; however, although the power of the earthquake in Haiti (magnitude 7.0) was much lower than that in Chile (magnitude 8.8), the loss of life in Haiti was far greater. An estimated 222,000 deaths resulted from the Haiti earthquake, as opposed to 521 in Chile.

SOURCE: USGS (2010).

ing as "a process by which community groups are helped to identify common problems or goals, mobilize resources, and in other ways develop and implement strategies for reaching the goals they collectively have set." Claudia Albano, a community advocate for the City of Oakland, California, defined community organizing as an approach that enables people, working together, to advance the cause of social justice.[2] She noted four community-organizing goals that contribute to the enhancement of community resilience, especially in communities that have other pressing issues: win concrete improvements in people's lives, empower people to speak and act effectively on their own behalf, effect institutional change, and develop an effective organization that wields the power of the community. Flexibility needed for sustainability can be partially achieved by allowing communities to determine their own priorities in addressing disaster and other community issues.

PRINCIPLES FOR SUCCESSFUL RESILIENCE-FOCUSED COLLABORATION

Previous sections of this chapter discussed the theoretical necessity of resilience-focused collaboration. This section begins the work of describing the theoretical basis for successful collaboration itself.

Identify and Create Incentives

Mandates and regulations are often seen by governments as the means to overcome barriers to collaboration and to provide incentives. For example, the 1986 Superfund Amendments[3] required communities to establish local emergency-planning committees consisting of representatives of chemical companies, public-safety agencies, and other organizations to protect the communities from the consequences of hazardous and toxic chemical contamination. Such legal requirements run the risk of forcing mere compliance or engendering only token, as opposed to substantive, collaboration. That point was discussed by participants in the committee's information-gathering workshop, especially in response to a presentation by Emily Walker regarding recommendations of the National Commission on Terrorist Attacks Upon the United States (also known as the 9-11 Commission)[4] for national standards for emergency preparedness and the establishment of an accreditation and certification program for business disaster resilience (NRC, 2010). American communities are extremely diverse in many dimensions, including population, geography, economic drivers, social and cultural factors, political climate, and civic infrastructure. This

[2]C. Albano, City of Oakland, Presentation to the committee, October 19, 2009.
[3]See www.epa.gov/superfund/policy/sara.htm (accessed March 12, 2010).
[4]See www.9-11commission.gov/ (accessed June 9, 2010).

tremendous variation warrants caution against mandating or prescribing a single approach to resilience-focused collaboration.

To start, collaboration succeeds when value is demonstrated and incentives are provided to participants in reaching communitywide goals. In commercial enterprise, the effect on the bottom line and return on investment are important incentives, but so is the ability to build trusted networks, to ensure better coordination with other community stakeholders, and to access information that enables accurate risk and benefit analyses and more effective business continuity planning. Participation may serve as good public relations for an organization, resulting in greater recognition of the organization's leadership in the community. As emphasized 45 years ago by economist Mancur Olson (1965), providing incentives for collaboration is especially challenging when collaborative activities aim at providing such public goods as environmental amenities, environmental quality in general, public health and safety, and disaster protection. Those are benefits that can be enjoyed in the future, even by those not involved in the efforts to achieve or preserve them. Incentives are essential to overcome tendencies of populations to let someone else solve problems.

Efforts to create collaboration focused on generating public goods often center on providing "selective incentives" that may be enjoyed only by those who agree to collaborate. Incentives that reduce the cost of joining collaborative relationships can be effective in overcoming "free riding," but the committee notes that what motivates small-business owner participation may not constitute a successful incentive for a faith-based organization or a branch office of a major corporation. Different strategies need to be devised to encourage participation of the full fabric of the community, including potential nondisaster-related benefits.

Ultimately, many participants will be motivated by enlightened self-interest, business continuity concerns, and the desire to serve the public good. Encouraging stakeholders to ask themselves questions such as *What will happen if we don't plan for a disaster?* and *Can we afford to not have the 'insurance' that investment in resilience provides?* may help guide them toward enlightened self-interest and participation in resilience-focused private–public collaboration.

Adopt an Appropriate Planning Perspective

The goals of collaboration will necessarily vary among communities because of differences in community priorities, vulnerabilities, culture, and resources. It is therefore impossible to design one model for collaboration that will be successful in all communities. Collaboration will most likely be successful if community resilience goals acknowledge the importance of identifying in advance the needs that will rise during each phase of the disaster cycle. Success of resilience-focused collaboration depends on planning in advance for disaster response and recovery. Adopting an appropriate planning perspective requires systematic identification of the resources and strategies needed to accommodate land-use planning, public preparedness education, and short- and long-term disaster recovery for likely scenarios. Flexibility in planning is vital because disasters do not follow plans. Incorporating flexibility into collaborative efforts will also allow communities to deal with unexpected disasters because networks and resources will already be in place. Although flexibility is a vital component of successful collaboration—and for resilience in general—collaborative relationships are more effective and sustainable if not created "on the fly" when a disaster occurs. On-the-fly relationships do not benefit from systematic planning and the bonds of trust such planning creates.

Agree on Decentralized vs. Centralized Decision Making

The capacity of communities to build disaster resilience is tied to how and how well all members of the community—individuals and organizations—engage in collaboration and benefit from outcomes. In the formative stages of collaboration, decisions that determine the roles and responsibilities of different participants in collaborative efforts are made. Forming, maintaining, and sustaining effective cross-sector relationships and implementing activities that are decided collectively are daunting but not impossible challenges. Centralized and decentralized organizational collaboration for disaster resilience present different merits.

Studies of real-world partnering activities have provided some insight into how collaboration can be organized, but there is no current research on disaster resilience-focused collaboration. Evaluations of communities involved in Project Impact (see Box 1.2) provide some information regarding effective organizational models for disaster resilience-focused collaboration. For example, the Disaster Research Center (DRC) at the University of Delaware evaluated seven Project Impact pilot communities and their networks with an emphasis on organizational and decision-making structures (e.g., Wachtendorf et al., 2002a). The evaluation found that local pilot programs exhibited varied centralized and decentralized decision-making structures and a variety of organizational structures, ranging from horizontal to hierarchic. The DRC also studied nonpilot Project Impact communities. In one study involving 10 communities of different sizes, the communities' organizational structures tended to be hierarchic and centralized, even though they organized their activities differently. That approach appeared to contribute to success in sustaining momentum in the period of time studied (Wachtendorf, 2002b).

The DRC stressed that most of the structures evolved—in response to goals, needs, and resources—into centralized structures (Wachtendorf, 2002a). Reports on those aspects of the program stressed that different organizational forms have both advantages and disadvantages. For example, tightly centralized collaborative structures offer the potential for better accountability, but they may also discourage innovation. The structure of Project

Impact networks also tended to shift as a consequence of project maturation, changes in focus, and mergers with other programs.

The committee recognizes that Project Impact was a short-lived program, and therefore the long-term benefits of one organizational structure over another cannot be determined from the evaluation of Project Impact communities. Additionally, Project Impact funding was provided to support a coordinating function in communities—some communities chose a hierarchical and centralized mechanism. Because a mechanism functioned for a period of time, and even functioned well, does not imply it was the "best" mechanism for that purpose or that it was sustainable. Given that, the committee turned to other sources of information on centralized versus decentralized approaches. For example:

- Economists have done much research related to optimizing incentives within organizations. Ján Zábojník studied the costs associated with centralized decision making (Zábojník, 2002). His research indicates that it could be more cost effective for an organization to allow its employees to decide the methods for doing their jobs—even if managers have better information—than it is to motivate employees to accept methods proscribed top-down. Employee morale is a factor in his calculations.
- In his examination of lessons learned from the private sector and U.S. Coast Guard responses to Hurricane Katrina, Steven Horwitz suggests that agencies with more decentralized structures (for example, the U.S. Coast Guard) were able to perform better following Hurricane Katrina than their more centralized counterparts in large part because they were more knowledgeable of the communities they were serving and because their decision making structure allowed them to respond more quickly to community needs. (Horwitz, 2008).
- In his discussion of organizational characteristics critical for successful disaster response, John Harrald described essential elements of organizing for and coordinating response to extreme events including a combination of discipline (in structure, doctrine, and process) and agility (in the ability to be creative, improvise, and adapt) (Harrald, 2006). Harrald describes research by several social scientists that confirms the necessity of adaptability, creativity, and improvisation in disaster response; that such is more likely in an environment where organizational learning and decision making are decentralized is possible.

Much of the literature reviewed by the committee described how organizations worked or had the potential to work together during disaster response, for example in response to Hurricane Katrina. Those examples strongly suggest that decentralized decision making within a structure is effective in disaster response. Whereas these examples are useful, they do not discuss how best to organize private–public collaboration during normal, nondisaster times, as encouraged in this report. Thus the nondisaster-related literature became important. The committee then considered relevant research literature, input received during its information-gathering workshop, and committee expertise. The committee concluded that an approach that emphasizes decentralized decision making and horizontal networks of collaborators—rather than top-down interactions between people and organizations—within a consensus-derived structural organization is best suited to achieving resilience goals. That conclusion was reached for several reasons. The horizontal network is the form of organization most compatible with the concept of collaboration. Just as collaborative arrangements aim at achieving goals once addressed by bureaucracies, networks can perform functions once performed by hierarchies. Although some may argue that centralization allows faster decision making and action, centralized organizations may be less effective in extremely stressful situations (e.g., Dynes, 2000). They have also been observed to be dependent on the skills, knowledge, and even personality of a core coordinator (Wachtendorf et al., 2002a).

In a related vein, network forms of organization are similar to the structure of the federal U.S. system of governance. Federalism is a decentralized form of governance that recognizes that different levels of government have distinctive resources and authorities and that public agencies at national, regional, state, and local levels develop their own distinctive types of collaborative arrangements.

Decentralized network arrangements are consistent with the growing importance of information as a force in contemporary societies and are well suited to information sharing in an "information society" that places a premium on knowledge management. Networks are increasingly prominent because of how societies and economies are organized. They are also consistent with the intent of the National Response Framework, which envisions a decentralized approach to disaster management and acknowledges local communities as the first line of defense when disasters strike. To consider decentralized decision making in collaborative arrangements and networks as a means of achieving resilience goals is therefore logical. The notion is further supported in the case of disaster management, in which gaining an awareness of local vulnerabilities, needs, and resources is paramount.

The committee cannot overstate the importance of community stakeholder agreement on the structure of collaboration and on decision-making processes before disaster strikes. Without agreement and "buy in" from the community on exactly how decisions will be made within the collaborative network, decisions—especially those made under stressful circumstances—may be met with resistance or distrust.

Allow for Multiple Levels of Engagement

Collaboration may occur in different forms and include different levels of engagement. It may occur through simple networking, through resource coordination, through information sharing, and through formal structural relationships. Simple networking requires the least commitment on the part of participants, and thus the least investment and risk, because organizations retain separate resources and authority. It involves only intermittent exchange of information, common awareness and understanding, and a common base of support (Butterfoss, 2007). More complex forms of networking may incorporate networking tools that allow systematic and sophisticated information exchange. Relationships are generally without clearly defined structure or mission but may involve cooperation on specific tasks. Entities may cooperate for any number of reasons, such as sharing information and avoiding duplication of effort.

Complex goals established for mutual benefit among participants require a greater degree of coordination between individuals or organizations and may result in more formal and longer-term relationships focused on specific tasks. Resources and rewards may be shared, but each organization retains separate resources and authority. The highest level of collaboration may include new structural arrangements and commitment to a common mission among all participants. Such arrangements are sometimes called partnerships or coalitions. Resources may be jointly secured or pooled, and results and rewards are shared. Power may or may not be equally shared, but all members generally have input into collaborative processes. Higher-level relationship such as these will not work unless trust and productivity levels are high.

Building community resilience involves sustained effort at all levels of collaboration. Different individuals, groups, and organizations contribute at different levels at any given time. The level of engagement in collaboration is contingent on willingness both to commit and to risk more in the interest of community resilience on the basis of perceived benefits to participants. As the level of engagement increases, linkages between organizations become more intense and more influenced by common goals, decisions, and rules and by resources participants make available. According to Winer and Ray (1994), collaboration changes the way organizations work together. Organizations move from competing to building consensus; from working alone to including others from diverse cultures, fields, and settings; from thinking mostly about activities, services, and programs to looking for complex, integrated interventions; and from focusing on short-term accomplishments to broad systems changes.

THE CONCEPTUAL MODEL

Conceptual models allow the user to visualize system elements and their relationships. In the same way a roadmap represents routes from one location to another, a conceptual model simplifies and abstracts a real-world system, depicts the probable causal relationships between system components, and helps to identify the true relationship between seemingly independent system elements (Sloman, 2005). Conceptual models are encouraged as a starting point for planning, for example, by the Substance Abuse and Mental Health Services

Administration of the U.S. Department of Health and Human Services in its guidance related to identifying and selecting evidence-based interventions (Center for Substance Abuse Prevention, 2009).

Strengthening community resilience through private–public engagement requires a conceptual framework that captures the unique characteristics of private–public collaboration. The committee developed a conceptual model for resilience-enhancing collaboration based on a community coalition action theory (CCAT) model for public-health applications developed by Butterfoss and Kegler (2002). Much of CCAT was borrowed from the fields of community development, community organization, citizen participation, community empowerment, political science, interorganizational relations, and group process (Butterfoss, 2007). Like the CCAT, the committee's model provides a theoretical basis for initiating, maintaining, and establishing as an accepted part of the culture the complex collaborative relationships needed to create disaster-resilient communities. The model is intended to be used by both practitioners (those focused primarily on community outcomes) and researchers (those interested in looking at individual model elements empirically).

The conceptual model (Figure 2.1) first considers how resilience-focused collaboration is formed so that it will be effective and sustainable. On the basis of input the committee received during its workshop and firsthand experience of committee members, sustainable collaboration is more likely if it is based on a bottom-up approach and acceptance of the need for collaboration. A realistic assessment of the community is necessary to identify the common issues, resources, and capacities that may be leveraged to greatest advantage to build resilience. Evaluating existing networks is an important part of the assessment (Milward and Provan, 2006). Methods and models for collaboration appropriate for the community may then be chosen that allow flexibility and creativity, but also include a neutral facilitating body to oversee collaborative activities, seek funding, and have other day-to-day operational roles. Once the structure is chosen and established, consistent effort is needed to make sure that the structure remains an accepted part of "doing community business." Collaboration itself is then best developed in stages that are revisited as new partners are recruited, plans are renewed, and missions, goals, and objectives are amended. Such tasks as recruiting and mobilizing members, refining the organizational structure, securing funding, building capacity, selecting and implementing strategies, evaluating outcomes, and refining strategies are best considered part of the normal functioning of collaborative efforts to ensure effectiveness and sustainability.

A conceptual model that accommodates the evolving and complex nature of comprehensive emergency-management systems is essential for developing resilience-focused collaboration, and the model has elements that can be applied to any collaborative network at any stage of development. Whereas disasters are common occurrences around the world, disasters in a given community can be high-consequence but low-probability events. Resilience-building collaboration requires constant maintenance to be effective. Regular assess-

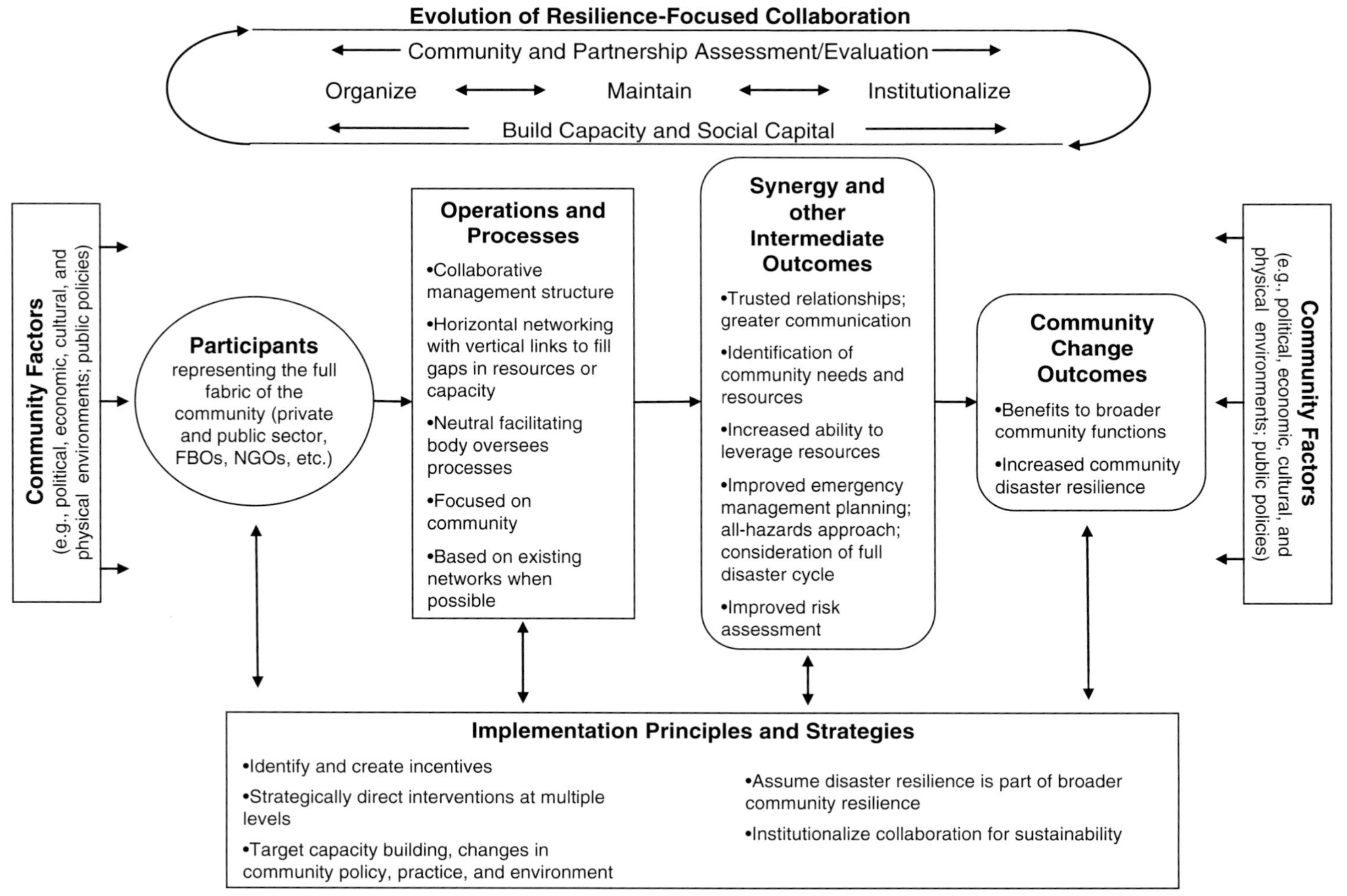

FIGURE 2.1 Conceptual model for private–public sector collaboration for building community resilience. Adapted from Butterfoss and Kegler (2002).

ment of the extent to which collaborative activities can or do result in favorable community change is a vital component of the committee's framework. Assessment of the collaborative structure, mission, objectives, and activities to maintain community relevance aids in the maintenance of collaboration and its acceptance by social institutions. Interdependences between critical infrastructure and networks need to be identified to create efficiency in the community (NRC, 2010). Interdependence and networking models will be effective if they are representative of current conditions, so regular reassessment is important. For those reasons, it is essential to understand, respond to, and even predict changes in the community in response to urbanization, shifting population densities, changing political administrations, and numerous other factors.

When assessing an existing collaborative network, organizers and participants may consider asking whether a bottom-up approach that ensures local-level acceptance and ownership in the scope and mission of the organization has been used. The assessment could also include such questions as: *Did collaboration reflect the array of issues that represented the community's concerns and interests? Are resources and capacities of the community understood so they can be leveraged to greatest advantage to build resilience? Are creative and flexible approaches applied in developing local resilience-building strategies?* The conceptual model may help organizers determine the correct questions to ask, as well as determine the answers to the questions. Those forming new resilience-focused collaborative structures can be guided by the conceptual model and a comparable set of questions.

Allowing collaboration to evolve in response to findings from regular assessments makes sustainability more likely. Because sustainability is more than a measure of financial stability, as noted during the committee's information-gathering workshop (NRC, 2010), assessments will be most useful if they measure sustainability of relationships in addition to financial and programmatic sustainability.

Model Components

The conceptual model (Figure 2.1) consists of six major nonlinear or sequential elements:

- *Community factors.* These are the external factors that constitute the input to planning during all stages of collaboration, such as jurisdictional challenges, the political climate, public policies, communication and trust between different levels of government or between agencies, and liability concerns. Other similar issues are geography, access to resources, current levels of community disaster readiness, trust of organized networks by the community, and understanding of terminology and jargon (Magsino, 2009). These are all factors that can affect participation in collaboration and the effectiveness of collaborative activities. Many of the factors, such as public policy, can themselves be influenced by collaborative action.

- *Participants.* Sustainable and effective resilience-focused collaboration depends on representation of the full fabric of the community. Private–public collaboration implies engagement between government entities; diverse industry sectors; NGOs, including faith-based, voluntary, and citizen organizations; and other elements of the community. The ability or inability of collaboration to respond to a threat is often decided by the composition of its members and their agency affiliations. According to workshop participants, failure to include the full fabric of the community, especially disenfranchised groups, can lead to an ineffective collaboration (NRC, 2010). A community-engagement approach uses strategies necessary to ensure that collaborators are equally vested in achieving collaboration goals. Other factors that contribute to successful recruitment and engagement of collaborators are the experience of members with the issue of concern and the involvement of community gatekeepers and members of groups that are diverse in their expertise, constituencies, sectors, perspectives, and background (Butterfoss, 2007).
- *Implementation principles and strategies.* A common understanding that disaster resilience is a part of broader community resilience is essential. Resilience-focused collaboration is successful when based on common goals and missions. Effective collaboration supports action strategies based on the resources and capacities available to the community. Efficient strategies are designed so that they are scalable and transferable to other collaborative and community efforts, regardless of the initial specific purpose. Interventions are more likely to build resilience if they include the entire community and if they are directed to different populations of the community in meaningful ways. Strategies to build capacity in all parts of the community to effect change in community policies, practice, and the environment are essential as are incentives to encourage and sustain participation in collaboration and community response to collaboration. Consideration of the sustainability of collaboration is vital in strategy planning. Sustainability is more likely if an understanding of the need for building resilience—and the need for private–public collaboration—is engendered in the community. Much as the business sector accepts the Chamber of Commerce as advocates for business concerns in the community, resilience-focused private–public collaborative structures will more likely be successful if accepted as advocates for overall community welfare.
- *Operations and processes.* These include the collaborative management structure, the various horizontal and vertical networking links within the structure, and a neutral convening or facilitating body to help organize collaborative activities and other day-to-day functions of collaboration, including recruiting and mobilizing members, securing funds, building capacity, selecting and implementing strategies,

evaluating outcomes, and refining strategies. Collaboration and leadership models are best when chosen on the basis of the needs and character of the community. An important role of leadership is breaking down or rendering more permeable any "silos" of interaction within networks that inhibit common cause, such as the emergency-management community working independently of the private sector. Processes to manage conflict appropriately; to weigh the costs and benefits of continuing participation, planning, and resources development; and to identify, save, and leverage community resources are all very important to success. Networks need not be built from scratch; efficiency is enhanced by recognition and incorporation of existing effective community organizational networks when feasible and consistent with collectively agreed on missions and goals. Careful design of collaboration structure and processes allows effective recruiting, capacity building, mobilization, securing of funding, selection and implementation strategies, evaluation of outcomes, and refinement of strategies to ensure effectiveness and sustainability.

- *Synergy and other intermediate outcomes.* Intermediate outcomes are beneficial results of the collaborative process, but not necessarily the final desired outcomes. They are the synergies created between organizations as a result of increased communication and trust, identification of community needs and resources, increased ability to leverage community resources for the good of the community, improved ability to assess community risks, and improved emergency and community management and planning. Stronger bonds between the private and public sectors are a result of collaboration. Those bonds will probably result in more effective assessment, planning, and implementation of all manner of community strategies (not just those for disaster resilience) and in tangible and intangible support and increased social networking within the community. With effective private–public collaboration comes an increased ability to resolve conflict within the community, a greater sense of belonging to the community, and a shared sense of local community ownership and responsibility among community members. The concept of collaboration synergy is predicated on the notion that individuals and organizations working together will accomplish more than could be accomplished by individuals separately.
- *Community change outcomes resulting in increased capacity and community disaster resilience.* These are changes in the community that increase community disaster resilience, such as changes in community policies, practice, and environment that result from enhanced community capacity and participation. Greater community resilience is evidenced by community organizations that can more effectively prepare for, respond to, and recover from disasters.

The nonlinearity of the model reflects the need for constant reassessment of the community and of the collaborative structure, goals, and strategies. As the community changes, it is in the community's best interest to reassess collaboration principles and strategies. This in turn triggers the necessity to evaluate the makeup of collaborating participants and the productivity of collaborative operations and processes. Peer mentoring—tapping into the expertise in other communities that have collaborated successfully—can be a valuable process for obtaining information on effective operations, processes, and strategies.

Much of the evidence supporting the validity of this conceptual framework and its guiding principles is anecdotal, and further examination of guidelines associated with the conceptual model is warranted. The conceptual model can be used by researchers as a roadmap to study and verify the systematic or logical connections of its elements, and determine, for example, metrics needed to assess the validity or progress of specific activities or outcomes. Ultimately, communities will adapt the framework according to their unique characteristics and locally determined issues and priorities. According to Mileti (1999: 63-64), "the process of transforming the future requires open-minded debate; full public participation; a willingness to experiment, learn, fine-tune, and alter; and a consensus among stakeholders to stand behind their shared commitment to the goal." That concept applies directly to communities attempting to build resilience as they identify and resolve gaps in knowledge and practice.

Chapter 3 of this report provides guidance on applying the concepts in the committee's conceptual model for private–public collaboration for enhancing community disaster resilience.

REFERENCES

Agranoff, R., and M. McGuire. 2003. *Collaborative Public Management: New Strategies for Local Governments*. Washington, DC: Georgetown University Press.

Burt, R. 2000. The Network of Social Capital. In *Research in Organizational Behavior*, R. Sutton and B. Staw, eds. Greenwich, CT: JAI Press, pp. 345-423.

Butterfoss, F. D. 2007. *Coalitions and Partnerships in Community Health*. San Francisco, CA: Jossey-Bass.

Butterfoss, F. D., and Kegler, M. C. 2002. Toward a comprehensive understanding of community coalitions: moving from practice to theory. In *Emerging Theories in Health Promotion Practice and Research*, eds. R. J. DiClemente, R. A. Crosby, and M. C. Kegler. San Francisco, CA: Jossey-Bass, pp. 157-193.

Center for Substance Abuse Prevention. 2009. Identifying and Selecting Evidence-Based Interventions Revised Guidance Document for the Strategic Prevention Framework State Incentive Grant Program. HHS Publication No. (SMA) 09-4205. Rockville, MD: U.S. Department of Health and Human Services. Available at prevention.samhsa.gov/evidencebased/evidencebased.pdf (accessed September 7, 2010).

DHS (Department of Homeland Security). 2009. National Infrastructure Protection Plan: Partnering to Enhance Protection and Resiliency. Washington, DC: U. S. Department of Homeland Security. Available at www.dhs.gov/xlibrary/assets/NIPP_Plan.pdf (accessed August 5, 2010).

Dynes, R. R. 2000. Government Systems for Disaster Management. Preliminary Paper No. 300. Newark, DE: University of Delaware Disaster Research Center. Available at dspace.udel.edu:8080/dspace/bitstream/handle/19716/672/PP300.pdf;jsessionid=571354606BF18BE02F43A1A1702831D1?sequence=1 (accessed September 8, 2010).

FEMA (Federal Emergency Management Agency). 2008. National Response Framework. Washington, DC: U.S. Department of Homeland Security. Available at www.fema.gov/pdf/emergency/nrf/nrf-core.pdf (accessed March 11, 2010).

Harrald, J. R. 2006. Agility and Discipline: Critical Success Factors for Disaster Response. *The Annals of the American Academy of Political and Social Science* 604(1): 256-272.

Heffner, R. C., ed. 2001. *Democracy in America*. New York: Penguin Books Ltd.

Horwitz, S. 2008. Making Hurricane Response More Effective: Lessons from the Private Sector and the Coast Guard during Katrina. *Mercatus Policy Series*, Policy Comment No. 17. Arlington, VA: Mercatus Center at George Mason University. Available at mercatus.org/uploadedFiles/Mercatus/Publications/PDF_20080319_MakingHurricaneReponseEffective.pdf (accessed September 8, 2010).

Magsino, S. 2009. *Applications of Social Network Analysis for Building Community Disaster Resilience: Workshop Summary*. Washington, DC: The National Academies Press.

McGuire, M. 2006. Collaborative Public Management: Assessing What We Know and How We Know It. *Public Administration Review* 66(1): 33-43.

Mileti, D. S. 1999. *Disasters by Design: A Reassessment of Natural Hazards in the United States*. Washington, DC: Joseph Henry Press.

Milward, H. B. and Provan, K. G. 2006. A Manager's Guide to Choosing and Using Collaborative Networks. *Networks and Partnerships Series*. Washington, DC: The IBM Center for the Business of Government.. Available at www.businessofgovernment.org/sites/default/files/CollaborativeNetworks.pdf (accessed September 2, 2010).

Minkler, M., and N. Wallerstein. 1999. Improving health through community organization and community building: A health education perspective. In *Community Organizing and Community Building for Health*, ed. M. Minkler. New Brunswick, NJ: Rutgers University Press.

Moynihan, D. 2005. Leveraging Collaborative Networks in Infrequent Emergency Situations. Washington, DC: IBM Center for the Business of Government.

NRC (National Research Council). 2010. *Private–public Sector Collaboration to Enhance Community Disaster Resilience: A Workshop Report*. Washington, DC: The National Academies Press.

Olson, M. 1965. *The Logic of Collective Action: Public Goods and the Theory of Groups*. Boston, MA: Harvard University Press.

Pickeral, T. 2005. Coalition Building and Democratic Principles. *Service-Learning Network* 11(1). Spring 2005 Constitutional Rights Foundation USA, Los Angeles, CA.

Rayner, S. 2006. Wicked Problems, Clumsy Solutions: Diagnoses and Prescriptions for Environmental Ills. Jack Beale Memorial Lecture on the Global Environment, University of New South Wales, Sydney, Australia. July 25. Available at www.ies.unsw.edu.au/events/jackBeale.htm (accessed March 12, 2010).

Rittel, H., and M.Webber. 1973. Dilemmas in a General Theory of Planning. *Policy Sciences* 4:155-169. Elsevier Scientific Publishing Company, Inc.: Amsterdam.

Sloman, S. A. 2005. Causal Models: How People Think about the World and its Alternatives. New York: Oxford University Press.

TISP (The Infrastructure Security Partnership). 2006. Regional Disaster Resilience: A Guide for Developing an Action Plan. Reston, VA: American Society of Civil Engineers. Available at www.tisp.org/tisp/file/rdr_guide[1].pdf (accessed March 12, 2010).

USGS (United States Geological Survey). 2010. 2010 Significant Earthquake and News Headlines Archive. Available at earthquake.usgs.gov/earthquakes/eqinthenews/ (accessed September 16, 2010).

Vigoda, E. 2002. From responsiveness to collaboration: Governance, citizens, and the next generation of public administration. *Public Administration Review* 62(5): 527-540.

Wachtendorf, T., R. Connell, B. Monahan, and K. J. Tierney. 2002a. Disaster Resistant Communities Initiative: Assessment of Ten Non-pilot Communities. Newark, DE: University of Delaware Disaster Research Center Final Project Report #48. Available at dspace.udel.edu:8080/dspace/handle/19716/1158 (accessed June 21, 2010).

Wachtendorf, T., R. Connell, K. Tierney, and K. Kompanik. 2002b. Disaster Resistant Communities Initiative: Assessment of the Pilot Phase—Year 3. Newark, DE: University of Delaware. Available at dspace.udel.edu:8080/dspace/handle/19716/1159 (accessed March 12, 2010).

Waugh, W. L. 2007. Testimony before the Subcommittee on Economic Development, Public Buildings, and Emergency Management, House Committee on Transportation and Infrastructure. September 11. Available at republicans.transportation.house.gov/Media/File/Testimony/EDPB/9-11-07-Waugh.pdf (accessed March 12, 2010).

Waugh, W. L., and G. Strieb. 2006. Collaboration and leadership for effective emergency management. *Public Administration Review* 66(1):131-140.

Winer, M., and K. Ray. 1994. *Collaboration Handbook: Creating, Sustaining and Enjoying the Journey.* Saint Paul, MN: Amherst H. Wilder Foundation.

Zábojník, J. 2002. Centralized and Decentralized Decision Making in Organizations. *Journal of Labor Economics.* 20(1): 1-22.

Guidelines for Community-Based Private–Public Collaboration

Effective resilience-focused private–public collaboration will often span geographic and political jurisdictions, include multiple agencies and levels of government, and cross other social, economic, and cultural boundaries. Collaborators recognize that no person or entity has all the expertise, insight, information, influence, or resources to build community resilience. Likewise, there will be impediments to collaboration that need to be recognized and addressed, including cultural, interpersonal, political, financial, and technical challenges. There is the barrier of physical separation caused by time and distance that cannot be completely offset, even by the most sophisticated communication technology. In sum, collaborative efforts are often complex. An organizing structure is therefore necessary to understand how the various components of collaboration relate to one another (Briggs et al., 2009). This chapter offers practical suggestions for applying the conceptual framework provided in Chapter 2.

Many different types of community actors mobilize to respond when disaster strikes. Postdisaster response networks are far larger and more complex than those envisioned in official disaster plans (NRC, 2006). For example, based on multiple data sources, Kapucu (2007) found that over 1,100 nonprofit organizations played some role in emergency response and postevent relief activities following the terrorist attacks of September 11, 2001, in New York City. Included in that number were nonprofit organizations that formed specifically to address the needs of those affected by the attacks. Also using multiple sources, Bevc (2010) identified more than 600 organizations whose activities focused directly on emergency response tasks such as search and rescue, fire suppression, and assisting victims and emergency workers. These organizations were involved in extensive networks of interaction and collaboration that emerged and evolved over time. The mobilization of a broad spectrum of community organizations and sectors is thus a key factor enabling effective disaster response. Response activities typically involve a formal or informal network, characterized by collaboration rather than command and control, with entities joining response networks to carry out activities that are deemed necessary, regardless of whether such activities are specified in plans. The committee cautions, however, that whatever role a collaborative network serves in the community, it should be consistent with and supportive of the legal authority of emergency management agencies.

As described in *Partnerships for Emergency Preparedness: Developing Partnerships* (LLIS, 2006), many communities' public-safety and private-sector entities have conducted planning and preparedness operations largely independent of one another. Few fully understand or appreciate the others' roles in emergency prevention, preparedness, mitigation, response, and recovery. Public-safety agencies often underestimate the private sector's interest and involvement in emergency-preparedness efforts. Private-sector groups overestimate the capabilities of government and fail to recognize the need for their own contributions to an incident response. In addition, the private sector often perceives cooperation with government agencies as risky because of the government's role in regulating their industries, concern about the protection of proprietary information, and the potential of legal liability.

It can be challenging to motivate private and public sectors to participate in resilience-focused collaboration that emphasizes a comprehensive management approach. How are organizations encouraged to plan for disaster mitigation and preparation, as well as response and recovery? How are organizations encouraged to do this collaboratively with others in their community? The committee describes engagement at the community level in the first major section of this chapter. In it, the importance of acknowledging local networks and network diversity are discussed, as are the importance of engaging needed expertise—either locally or further afield—and following evidence-based principles of emergency management. The second major section explores structure and process in resilience-related activities, including the importance of a coordinating function and multilevel relationships. The third major section of the chapter discusses practical application of the conceptual model discussed in Chapter 2. The final section of this chapter provides the committee's overarching guidelines designed to address community-level private–public collaboration for enhancing disaster resilience.

ENGAGING AT THE COMMUNITY LEVEL

Just as there is no clear federal coordination or national strategy for climate adaptation (NRC, 2010a), there is no national strategy for building community disaster resilience. That 2010 NRC report on climate change concludes that there is a need for a national strategy for climate adaptation, and that the strategy would benefit from "a 'bottom-up' approach that builds on and supports existing efforts and experiences" at the state and local levels, including private–public collaboration.

This report does not address all components of a national resilience strategy, but the committee recognizes that with or without a national strategy, there is a need for community-level resilience. Achieving resilience at the state or national levels begins with resilience-enhancing efforts in local communities. Community efforts begin with individuals from any sector believing in and acting on a sense of personal responsibility to ensure community sustainability. Those individuals also convince others of similar need to act. Leadership and initiative can come from any sector.

Local government and local business and civic organizations have unique knowledge of, access to, and communication with individual citizens throughout the community. Well-prepared individuals contribute to household and workplace resilience. Well-prepared households and businesses contribute to neighborhood, social, commercial, economic, and community resilience. Well-prepared communities place fewer demands on state and federal resources because they are better able to cope when disasters or other disruptions occur. A nation is resilient when it is made up of resilient communities.

The notion that disaster resilience is fostered at the local community level is a cornerstone of many recent national preparedness efforts, including those of the Department of Homeland Security (DHS) National Response Framework. It states in part that "an effective, unified national response requires layered, mutually supporting capabilities" (FEMA, 2008:4), and that "resilient communities begin with prepared individuals and depend on the leadership and engagement of local government, NGOs, and the private sector" (FEMA, 2008:5). The concept of a "tiered response," a key element of the framework, places primary responsibility for hazard and disaster management at the local community level. Although it indicates that response activities must be flexible and scalable, the framework contains the directive that "incidents must be managed at the lowest possible jurisdictional level and supported by additional capabilities when needed" and states further that "incidents begin and end locally, and most are wholly managed at the local level" (FEMA, 2008:10). It can even be hypothesized, as Mileti (1999) did, that an indicator of community disaster resilience is the ability of a local community to cope with events without relying excessively on outside resources. Conversely, as seen during the January 2010 Haiti earthquake, communities and societies that lack disaster resilience may depend almost exclusively on external aid.

However, community and extracommunity preparedness efforts aid and reinforce household, business, and individual preparedness. Community resilience-enhancing interventions can thus be used at any level of analysis—individuals, households, neighborhoods and community associations, individual businesses and groups of businesses, individual nonprofit organizations and networks of nonprofit organizations—with a key stipulation that such efforts be mutually reinforcing.

The sections that follow discuss the strategic dimensions of a national framework for enhancing disaster resilience with an emphasis on local-level strategies. The committee was asked to focus on community-level private–public collaboration, and it did, but the committee would be remiss to ignore the sociopolitical environment that is conducive to such collaboration. Discussions of strategy are based on what has been learned not only in the fields of emergency management and disaster-loss reduction but in other fields such as public health. The committee also draws on knowledge on topics such as citizen mobilization, collective action, and community organizing.

Recognize the Significance of Local Networks

Chapter 1 defined a community as a group of people with a common interest, on the basis of the definition developed by Etienne Wenger (1998). The concept of community has many dimensions, and communities are perhaps best understood as consisting of networks of relationships of various types on various scales. Networks exist at many levels within and across myriad sectors of society, including interpersonal, neighborhood, organizational, private industry, civic, and governmental. Networks may be based on informal or formal ties or on a mixture of the two. They may be organized according to geography, government or economic functions, or interests of various kinds. Communities in the United States include a wide array of vibrant and dynamic networks, and even the most impoverished and seemingly deprived communities and subsets of communities include such networks. U.S. society is widely understood to contain a rich array of religious institutions, voluntary associations, nonprofits, coalitions, interest groups, and alliances of other kinds.

Efforts to mobilize individuals and groups are most efficient and successful when begun through existing networks and institutions using multiple mechanisms. People are not motivated to work toward a goal as isolated individuals; rather, they participate in collective efforts through the groups and institutions in which they normally participate. In the U.S. civil rights movement, for example, black churches and church-related networks, such as the Southern Christian Leadership Conference, provided a means of connecting individuals to the movement (Morris, 1984). Mass-media–based information campaigns, such the DHS Web portal,[1] may succeed in bringing issues to the attention of individuals but might not be effective in motivating collective action.

An individual business owner may understand that preparing for disasters is important but might not act on that understanding unless messages and encouragement come through the local chamber of commerce or other business-oriented association. That was the experience of the Disaster Resistant Business Toolkit (DRB Toolkit)[2] Workgroup that brought together private- and public-sector experts in business continuity and emergency management. Through existing relationships, the workgroup developed and launched disaster planning software to assist small businesses in the United States with continuity planning to reduce their vulnerability to all hazards. The DRB Toolkit Workgroup understood the interconnectedness of a community (Bullock et al., 2009).

All-inclusive networks can be created by linking and optimizing existing professional,

[1]Available at www.ready.gov/ (accessed July 1, 2010).
[2]See www.DRBToolkit.org/ (accessed July 30, 2010).

religious, service, social, economic, and other networks. Community stakeholders collaborating in resilience-enhancing strategies might therefore consider how to reach individuals and groups through the organizations to which they belong. It is important to build on the work and achievements of local networks devoted specifically to emergency management and homeland security, such as local disaster-preparedness networks and the DHS Urban Areas Security Initiative programs that can provide a firm basis for more inclusive and comprehensive resilience-enhancing efforts. However, it is equally important for resilience-enhancing efforts to be directed toward and occur through the wide array of local entities, associations, and alliances that represent the full fabric of the community. One example of a network of community-based organizations is the Kentucky Outreach and Information Network (KOIN),[3] established to communicate with hard-to-reach populations in an emergency. KOIN is a network of local resources that provides information to groups such as non-English speakers and the deaf, and its members serve as liaisons between those people and emergency responders.

Collaboration with local agencies can increase the effectiveness of collaboration, not only because of increased interaction with the emergency management community, but because of the relationships of local organizations with members of the community. Local police and fire departments, for example, have relationships with citizen groups such as neighborhood crime watch groups, Community Emergency Response Teams,[4] or Citizens of Oakland Respond to Emergencies (CORE) in Oakland, California.[5] Engaging with groups such as local Boys and Girls Clubs,[6] for example, or groups representing minorities, or special needs groups in a community may help reach individuals who are otherwise difficult to reach. Most groups will have their own trusted means of communication, and people respond best to information that comes from people they know and trust and with whom they interact regularly. That is the case regardless of the type of information conveyed, as was demonstrated through examples in the summary of a National Research Council workshop on social-network analysis for enhancing community resilience (Magsino, 2009). As noted in the earlier National Research Council report on the human dimensions of hazards and disasters (NRC, 2006), horizontal ties both elicit and increase trust.

Recognize Network Diversity

Participants of the committee's information-gathering workshop noted the tendency of government to focus on "generic" populations—for example middle-class, educated suburban dwellers—that may not represent the diversity in the community or its networks.

[3]See chfs.ky.gov/dph/epi/preparedness/KOIN.htm (accessed September 15, 2010).
[4]See www.citizencorps.gov/cert/ (accessed September 15, 2010).
[5]See www.oaklandnet.com/fire/core/about.html (accessed September 15, 2010).
[6]See www.bgca.org/Pages/index.aspx (accessed September 15, 2010).

Engaging all sectors, community members, and existing networks increases the ability to identify community needs and leverage community resources. However, different communication mechanisms may be needed to communicate collaborative goals, functions, and benefits for different constituencies. Such tools may include conceptual models, narrative descriptions, and business prospectuses. In some cases, the mechanisms may need to be provided in different languages.

Successful use of existing networks includes recognition that not only do communities consist of numerous diverse networks, but that networks that include some community members by definition exclude others. Church membership, for example, is an important community tie for many people, but such networks are themselves diverse, and communities contain many people who are not church goers or affiliated with any religious institution. In most communities, there are well-established fraternal associations, but they are also diverse. Chambers of commerce and such institutions as the United Way serve as focal points for many—not all—local businesses and nonprofits, respectively. Similarly, many—not all—communities have a wide diversity of neighborhood and homeowners associations. There are also work-based and school-based networks. In some communities, a major employer provides a focal point for community activities. In our culturally diverse society, many networks center on ethnic identities, immigration histories, and minority community institutions. Ethnic enclaves have their own distinctive forms of social organization, which may not be well understood by the larger majority community.

The committee calls attention to other types of organizations, those that emerge in response to crisis. Participants of the committee's information-gathering workshop recognized that it is often through informal or unofficial channels that food, shelter, hygiene, and other support services are first offered immediately following a disaster (NRC, 2010b). Often groups, called "problem solving networks" by Milward and Provan (2006), emerge specifically to determine a way to quickly resolve the crisis and can result in long-lived and effective networks. Groups that arise in response to crisis are often not recognized or used effectively by emergency management officials. This report emphasizes the importance of preparation prior to a crisis and does not focus on groups that arise as a result of crisis. However, regular assessment of networks within a community may help identify the conditions under which such groups emerge. Mothers Against Drunk Driving,[7] for example, arose in response to tragic events in individual families, but is now a national nonprofit organization that promotes change in social behavior and in public policy. Understanding how such groups emerge may help communities understand where they may emerge, as well as how they may be used during crisis.

Private–public collaboration—whether directed at enhancing a community's quality of life, solving community problems, or, in this case, aiding communities in becoming disaster-

[7]See www.madd.org/About-Us/About-Us/Mission-Statement.aspx (accessed September 3, 2010).

resilient—will be most successful when it includes an early comprehensive assessment of diverse community network assets.

Engage Expertise at Local and Broader Scales

Different types of expertise are required for the development and maintenance of resilience-focused private–public collaboration. It is necessary for partners in such efforts to understand community risks, hazards, and vulnerabilities. Different kinds of information are required to address those needs, including hazard assessments, information on the impacts of past disasters, and information on the vulnerability of population groupings, the built environment, and ecosystems. Community stakeholders require a general understanding of such issues as how much protection current building codes offer against damage, the likely consequences of previous land-use decisions, and the likely social and economic impacts of both probable and worst-case disaster events. Those kinds of information can come from multiple sources, including loss-estimation studies that use HAZUS,[8] HAZUS-MH, the Social Vulnerability Index (SOVI),[9] census data, community disaster scenarios, and individuals and organizations including university researchers, professional engineers and engineering societies, building-code officials, urban planners, and state agencies. Resilience initiatives may also draw on the knowledge of community-based experts, such as community organizers, elected and appointed officials, leaders in community-based nonprofit organizations and businesses, and long-term community residents. Such information lends nuance and meaning to more "scientific" hazard and vulnerability data and increases the probability that resilience-enhancing private–public collaboration will be successful.

With support from FEMA, the National Science Foundation–sponsored Multidisciplinary Center for Earthquake Engineering Research developed a set of guidelines for seismic-safety advocacy strategies (Alesch et al., 2004). The report contains practical guidance on a variety of topics, including how to use scientific expertise in community loss-reduction campaigns, risk communication, community mobilization, and partnership building. Although the report is focused on earthquake safety, its lessons are easily transferable to an all-hazards context. Individual community stakeholders are not all expected to be able to identify what information is available or to determine what actions are translatable or scalable to their own circumstances. Private–public collaboration therefore benefits greatly from engaging those that have necessary expertise—for example, from local institutions of higher learning—as community stakeholders.

[8]See www.fema.gov/plan/prevent/hazus/ (accessed July 1, 2010).
[9]See webra.cas.sc.edu/hvri/products/sovi.aspx (accessed July 1, 2010).

Follow Evidence-Based Principles of Community Engagement

Community engagement is a well-recognized approach to community problem solving that has been used in such fields as health care and research, law enforcement, and planning for pandemic influenza and homeland security threats (Patterson et al., 2010; Fleischman, 2007; NRC, 2006; Lasker et al., 2003). Numerous resources exist for those desiring to engage the full fabric of the community in community activities. The Higher Education Network for Community Engagement is made up of community colleges, colleges, and universities that provide community-engagement guidance.[10] Numerous online resources contain step-by-step guidelines on effective community-engagement processes. Such organizations as the Center for Advances in Public Engagement[11] provide an array of materials that can inform local resilience-enhancing efforts, including what the center terms "core principles of community engagement" (Kadlec and Friedman, 2008). The IBM Center for the Business of Government[12] offers an online *Collaboration Series* that includes guidance for public managers involved with citizen engagement (Lukensmeyer and Torres, 2006). A report on the promises and challenges of neighborhood-level democracy, based on a meeting organized by Grassroots Grantmakers and the Deliberative Democracy Consortium, explores creative ways for local governments to engage citizens in public decision making and problem solving (Leighninger, 2009).

New initiatives also seek to apply concepts of community engagement originally developed in the fields of health and public health to disaster preparedness. For example, the National Resource Center on Advancing Emergency Preparedness for Culturally Diverse Communities is a project of the Drexel University School of Public Health Center for Health Equality that seeks to link health-based and disaster-loss reduction engagement strategies.[13]

In 1995, the Centers for Disease Control and Disease Prevention and the Agency for Toxic Substances and Disease Registry established a Committee for Community Engagement, which reviewed relevant research and synthesized findings in a report titled *Principles of Community Engagement* (CDC-ATSDR, 1997). The recommendations in that report are applicable to all types of community-based improvement efforts, including resilience initiatives, and are summarized in Box 3.1.

[10] See www.henceonline.org/ (accessed June 30, 2010).
[11] See www.publicagenda.org/cape (accessed June 30, 2010).
[12] See www.businessofgovernment.org (accessed August 31, 2010).
[13] See www.diversitypreparedness.org/ (accessed June 30, 2010).

BOX 3.1
Principles of Community Engagement as Recommended by the CDC-ATSDR Committee on Community Engagement

The Centers for Disease Control and Prevention established the Committee for Community Engagement in 1995 to consider the literature and practical experience gained by those who were engaging people and organizations in communities around the nation and to provide public-health professionals and community leaders with scientific information and practical guidelines to aid in decision making and action on issues associated with health promotion, health protection, and disease prevention. The community engagement strategies provide practical guidance for those wishing to engage in resilience-focused private–public collaboration. The following is a summary of strategies drawn directly from that committee's report (CDC-ATSDR, 1997).

Before Beginning a Community Engagement Effort

1. Be clear about the purposes or goals of the engagement effort, and the populations and communities you want to engage.
2. Become knowledgeable about the community in terms of its economic conditions, political structures, norms and values, demographic trends, history, and experience with engagement efforts. Learn about the community's perceptions of those initiating the engagement activities.

For Engagement to Occur

3. Go into the community, establish relationships, build trust, work with the formal and informal leadership, and seek commitment from community organizations and leaders to create processes for mobilizing the community.
4. Remember and accept that community self-determination is the responsibility and right of all people who comprise a community. No external entity should assume it can bestow on a community the power to act in its own self-interest.

For Engagement to Succeed

5. [Partner] with the community . . . to create change and improve health [and resilience].
6. All aspects of community engagement must recognize and respect community diversity. Awareness of the various cultures of a community and other factors of diversity must be paramount in designing and implementing community engagement approaches.
7. Community engagement can only be sustained by identifying and mobilizing community assets, and by developing capacities and resources for community health decisions and action.
8. An engaging organization or individual change agent must be prepared to release control of actions or interventions to the community, and be flexible enough to meet the changing needs of the community.
9. Community collaboration requires long-term commitment by the engaging organization and its partners.

SOURCE: CDC-ATSDR (1997).

STRUCTURE AND PROCESS IN RESILIENCE-RELATED ACTIVITIES

Collaboration to achieve disaster resilience requires considerable attention to organizational design and structure. Insufficient attention to organization is likely to result in short-lived partnerships that fail to achieve their objectives. Inappropriate forms of organization can lead to participant dissatisfaction and conflict among stakeholders. Accordingly, the committee gathered research-based evidence on appropriate forms of organization for collaborative networks and collected the views of experts regarding best practices. The committee's conceptual model for resilience-focused private–public collaboration (Figure 2.1) can serve as a visual reminder of the connections between various collaborative elements and desired outcomes. Referring to the conceptual model while planning and mobilizing a collaborative network can assist organizers in decision making and assessment of activities.

The Importance of a Coordinating Function

The University of Delaware Disaster Research Center Project Impact assessment studies emphasized the importance of local Project Impact coordinators, whose jobs consisted of ensuring that communities were progressing in collaboration, partnership building, and other project goals. The findings suggest that regardless of how collaborative activities are organized, it is necessary to devote resources specifically for collaboration management. Put another way, it appears to be insufficient to argue for the importance of collaboration without also investing in individuals or groups that are charged with the responsibility of ensuring that collaboration is taking place. The experience of dedicated staff ultimately reduces jurisdictional confusion and wrangling after a disaster, allows more efficient pooling of resources, and promotes faster recovery. It is relatively easy to persuade potential collaborators to join umbrella organizations or to be signatories to disaster plans. However, given the infrequency of serious disasters in any given community, it is far more challenging to engage their active participation in resilience efforts on an ongoing basis. A strong collaborative network with dedicated staff will help keep loss reduction and resilience a community priority as an integral part of normal community functioning.

Some may argue that a coordinating function is not consistent with the committee's suggestion that decision making remain decentralized. The committee would counter that decentralized decision making is possible within an organized structure. Our system of governance in this country is an example. Rules and guidelines exist to direct the structure, but the structure does not direct the outcomes of decision-making processes. As long as there is consensus regarding rules of collaboration and the actions of a coordinating person or body, and as long as those rules are regularly evaluated for their relevance, decentralized decision making is possible.

Communities may decide that resources are too scarce to support a dedicated coordinator; it is essential that they consider the greater cost of *not* having a coordinator and the long-term benefits a coordinator is likely to provide.

Horizontal and Vertical Ties

Because enhancing disaster resilience is a nationwide goal, it is most useful to consider collaborative activities in the context of individual efforts nationwide. That does not imply that collaborative efforts should be driven by federal regulations and requirements or that collaboration should be approached in a uniform fashion in communities around the country. As with any program designed to address national problems, successful solutions developed to improve disaster resilience reflect the diversity of local communities around the nation and are consistent with the structure of the U.S. intergovernment system. Because of the importance of local-level buy-in to sustain the effort, it can be counterproductive for higher organizational levels in both the private and public sectors to provide more than technical, logistical, or financial support unless requested and coordinated with local leadership.

Chapter 2 discussed the importance of developing strong horizontal or intracommunity networks for disaster resilience. The conceptual model (Figure 2.1) includes strategizing for including the full fabric of the community. It is appropriate that horizontal networks receive substantial emphasis, but ideally resilience-enhancing programs will include a productive mix of horizontal and vertical collaboration and coordination. In its 2006 report *Facing Hazards and Disasters: Understanding Human Dimensions,* the National Research Council linked disaster resilience to the concept of social capital and emphasized the importance of both horizontal integration (within the community) and vertical integration (across different scales) among entities participating in loss-reduction activities. Regarding the importance of strong horizontal ties, the report (NRC, 2006: 231) noted that a community with a high degree of horizontal integration (i.e., strong social capital) has an active civic engagement program that fosters more tightly knit social networks among citizens and local organizations. Stronger networks provide a greater opportunity for creating interpersonal trust. Such a community can be a viable, locally based problem-solving entity. Its organizations and individuals not only have an interest in solving public problems, but also tend to have frequent and sustained interaction, believe in one another, and work together to build consensus and act collectively. Thus, local populations have the opportunity to define and communicate their needs, mediate disagreements, and participate in local organizational decision making.

Intracommunity ties thus constitute the fundamental building blocks of a disaster-resilient society. However, there is also the need to link communities vertically to other external entities. External ties—for example, among local communities and state and federal governments; local companies and their parent corporations; and local chapters of nonprofits and their national headquarters—bring benefits that cannot be realized through intracommunity linkages alone. The benefits include connections to broader societal institutions, expansion of trusted networks, and greater access to funding, expertise, and other resources (NRC, 2006). Both types of integration—intracommunity ties and external ties—are necessary to maximize the ability of communities to mobilize, learn, and innovate.

Emphasizing the importance of vertical ties between community networks and external entities does not imply that communities relinquish their decision-making authority to outside control. Rather, in keeping with the spirit of this entire report, collaboration and partnering are the most appropriate forms of interaction across scales for several reasons. First, mandates and regulations inevitably encounter resistance. Second, in the case of disasters and in many other situations, external entities that provide such resources as information and funding in fact have no formal authority over the vast majority of local network actors. For example, the federal government cannot require a local community to adopt a particular building code; mandate that corporations adopt *NFPA 1600*, the National Fire Protection Association *Standard on Disaster/Emergency Management and Business Continuity*;[14] require small businesses to prepare for disasters; force communities to tax themselves to achieve higher levels of resilience; or regulate the disaster-relevant activities of local nonprofits. Third, community residents and community-based entities have local knowledge that is not available to entities at other levels of scale. For example, in an increasingly global society, large corporations may lack a detailed understanding of the hazards faced by their local affiliates.

Project Impact provides a useful example of federal leadership not accompanied by efforts to control or micromanage local disaster-loss reduction activities. The Federal Emergency Management Agency (FEMA) provided financial assistance directly to local communities to engage in four types of activities: hazard assessment, hazard mitigation, public education, and the formation of private–public partnerships. FEMA provided general guidelines to participating communities and encouraged their development of memoranda of understanding to formalize partnerships related to loss reduction, but it did not tell communities what to do or how to work toward the four goals, nor did it make funding conditional on adopting particular types of organizational forms or processes. Government policy can nourish or diminish resilience. Federal policy that links the provision of resources to top-down control, federally mandated priorities, or uniform implementation has the potential to reduce the flexibility, innovation, and capability of individual communities in engaging and sustaining nongovernment stakeholders in resiliency efforts.

Parallels exist between efforts to achieve disaster resilience and efforts to respond to climate change and variability. Both involve the management of risks, and in both cases local

[14]See www.nfpa.org/aboutthecodes/AboutTheCodes.asp?DocNum=1600 (accessed July 1, 2010).

communities, states, and regions are engaging in innovative approaches that often extend beyond what the federal government requires. With respect to climate change, the National Research Council America's Climate Choices project report on informing climate-related decisions emphasizes that the federal government should not attempt to preempt local climate-change mitigation and adaptation initiatives or stifle innovative programs (NRC, 2010a). But the report also indicates that the federal government can do much, particularly in providing information. The same may well be true for encouraging disaster-resilient communities, and many of the collaboration models and case studies provided in the climate change report can be useful in the context of disaster resilience-focused private–public collaboration. With these ideas in mind, communities can establish vertical ties and collaborative relationships with state- and national-level organizations and governments.

BUILDING AND OPERATING COLLABORATIVE PARTNERSHIPS: PRACTICAL APPLICATION OF THE CONCEPTUAL MODEL

Many local communities, business and professional organizations, and state and local agencies have broken new ground in creating self-governed, private–public partnerships to serve their constituencies, and examples of these are provided throughout this report. These partnerships provide a promising array of collaboration models and lessons learned. The committee recognizes the need for a national framework that will enable the development of community-based partnerships; however, until such a framework exists—with supporting policy and resources—there is intrinsic value in developing entrepreneurial partnership work at the state and community levels.

The benefits of establishing community and statewide private–public collaboration before a disaster strikes have been observed in recent years. Those seeking to build private–public collaboration in their own communities may wish to use existing efforts as models. For example, the Safeguard Iowa Partnership[15] played a pivotal role during response to the historic 2008 floods in the midwestern United States; the Earthquake Country Alliance[16] staged annual statewide earthquake drills in California[17] and supports earthquake preparedness in multiple states and some other countries; the Aware and Prepare Program in Santa Barbara, California, is a private–public collaboration established by the Orfalea Foundation to increase the level of community disaster preparedness;[18] and regional alliances have expanded their collaboration on economic issues to include disaster resilience.[19] The

[15]See www.safeguardiowa.org (accessed July 1, 2010).
[16]See www.earthquakecountry.org/ (accessed July 30, 2010).
[17]See www.ShakeOut.org/ (accessed July 30, 2010).
[18]See www.orfaleafoundations.org/go/our-initiatives/aware-prepare/ (accessed July 1, 2010).
[19]See www.pnwer.org (accessed July 1, 2010).

committee heard examples of a variety of other collaborative efforts during its information-gathering workshop (NRC, 2010b).

Although there is anecdotal evidence of success and growing support for the concept of private–public collaboration, the expertise and resources required to sustain all-hazards community partnerships on a nationwide basis are lacking, and hundreds of state and local agencies, private businesses, and NGOs are looking for guidance in the "how-tos" of collaboration. The committee considered numerous examples and case studies of resilience-focused private–public collaboration and identified common strategies among them for developing effective communitywide collaboration. For example, models developed by the Michigan State University Critical Incident Protocol (MSU-CIP) Community Facilitation Program[20] and by Business Executives for National Security (BENS)[21] have been applied in diverse communities nationwide. Other collaboration models and steps recommended for collaboration implementation are referred to in the series *Public Private Partnerships for Emergency Preparedness*, published on Lessons Learned Information Sharing (LLIS).[22] The federally funded Community and Regional Resiliency Institute has also demonstrated effective partnership development through its three pilot communities—in Charleston, South Carolina; Memphis, Tennessee; and Gulfport, Mississippi.[23]

Resilience-focused collaboration attempts to build social capital in the community. A good source of discussion on this and other types of networks can be found in a document by Milward and Provan (2006). They describe essential management tasks in public networks that can be adapted and applied to private–public collaboration. A modified version of their table is provided as Table 3.1. The committee finds that those tasks are consistent with the application of its conceptual model (Figure 2.1).

In the next sections, the committee describes the developmental steps it considers most common and effective for private–public collaboration born from grassroots efforts. The conceptual model (Figure 2.1) can be a valuable tool when applying these suggested steps. Those collaborating may decide, based on the conceptual model, that certain aspects of their collaboration warrant change to get the best outcomes, or the model can be modified as collaboration and the community change over time. The conceptual model can be consulted regarding many decisions about structure, processes, strategies, and desired collaboration outcomes. Options can be compared to the model to determine which is most consistent with collaborative goals.

[20]See www.cip.msu.edu/ (accessed July 1, 2010).
[21]See www.bens.org/ (accessed July 1, 2010).
[22]See www.llis.gov (accessed July 1, 2010).
[23]See www.resilientus.org (accessed July 1, 2010).

Identify Leadership

Community-based private–public collaboration often begins with the inspiration of an individual leader—a business leader, state or local government official, civic-minded community organizer, or public servant—who sees the value of building an alliance to address a particular need. That person may already be an established leader in the community but could also be a concerned citizen who builds support and buy-in at the grassroots level. The person may not envision a communitywide collaboration to address myriad issues, but his or her initial outreach begins the collaborative process and lays the foundation for a broader, more inclusive partnership. Initial goals could include creating an advisory or leadership team.

Create an Advisory or Leadership Team

A small core team of three to six champions is ideal for beginning to frame the general goals of the collaborative effort and for exploring potential opportunities and benefits. The person who initiates collaboration—described above—may form and be part of the team, or may ask for guidance from others who could then form the team. A main function of the core team is to define the general purpose of collaboration before broader participation is invited, leaving specific functions open to discussion during the early stages of development. The latter is necessary to build consensus and buy-in among key stakeholders. The most effective core team is one that is representative of the community at various levels.

It is during this early exploratory phase that relevant top public officials and high-level private-sector leaders may be approached to enlist champions from both the public and private sectors (MSU, 2000). If a collaborative effort is directed solely by a government agency, there is the risk that businesses and other nongovernment stakeholders will view the effort as "just another government program." Conversely, public officials will be more likely to support a collaborative initiative spearheaded from within the private sector or by a private citizen if they are brought into the development process early.

Using a conceptual model such as provided in Figure 2.1 will help the core team to determine a preliminary framework for their own collaborative network and help the core team keep the appropriate goals in mind as collaboration expands.

Invite Key Stakeholders to the Table

The size and breadth of a collaborative relationship will be determined by its scope and mission, which may expand as collaboration matures. It is essential that the core advisory group developing the collaborative effort begin to identify other constituencies to be included in later stages of development. Convening too large a group at the outset may prevent

TABLE 3.1 Essential Management Tasks in Collaboration

Essential Networks Management Tasks	Management *of* Collaboration	Management *in* Collaboration
Management of Accountability	• Determining who is responsible for which outcomes. • Rewarding and reinforcing compliance with collaboration goals. • Monitoring and responding to collaboration "free riders."	• Monitoring your organization's involvement in collaboration. • Ensuring that dedicated resources are actually used for collaborative activities. • Ensuring that your organization gets credit for contributions to collaboration. • Resisting efforts to "free ride."
Management of Legitimacy	• Building and maintaining legitimacy of the collaborative concept, structures, and involvement. • Attracting positive publicity, resources, new members, tangible successes, etc.	• Demonstrating to others (members, stakeholders) the value of participation in collaboration. • Legitimizing the role of the member organization among other collaborators.
Management of Conflict	• Setting up mechanisms for conflict and dispute resolution. • Acting as a "good faith" broker. • Making decisions that reflect collaboration-wide goals and not the specific interests of members.	• Working to avoid and resolve problems with individual collaborators. • Working inside your organization to act as a "linking pin" to balance member organization versus collaboration demands and needs.
Management of Collaborative Structure	• Determining which structural models would be most appropriate for the success of collaboration. • Implementing and managing the structure. • Recognizing when structure should change based on collaboration and participant needs.	• Working effectively with other collaborators and with collaborative management, based on the structure of collaboration in place. • Accepting some loss of control over collaborative decisions.

TABLE 3.1 Continued

Essential Networks Management Tasks	Management *of* Collaboration	Management *in* Collaboration
Management of Commitment	• Getting the "buy-in" of participants. • Working with participants to ensure they understand how success of collaboration can contribute to the organization's effectiveness. • Ensuring that collaborative resources are distributed equitably to network participants based on collaborative needs. • Ensuring that participants are well informed about collaborative activities.	• Building commitment within the member organization to goals of collaboration. • Institutionalizing involvement in collaboration so that support of collaboration goals and participation goes beyond a single person in the organization.

SOURCE: Milward and Provan (2006).

effective relationships from forming and make self-governing impossible. An example of a specialized private–public partnership that engages only a particular specialized community is the Twin Cities Security Partnership, which was developed to increase public safety and quality of life in the Minneapolis area. The private sector collaborates with law-enforcement officials to share intelligence, threat alerts and warnings, and the potential for security incidents on a regular basis.[24] Collaboration began among a core group of business leaders and law-enforcement officials in 2003 and now has more than 100 members. An applicant for membership, however, must be a security practitioner, a supplier of security service, a management or common-level law-enforcement official, or a critical-infrastructure official. In this particular partnership, the key stakeholders are those most familiar with the issues associated with security.

Expertise needs will be broader in the case of disaster-focused and community resilience-focused private–public collaboration that reaches the full fabric of the community. Identifying key stakeholders who have the necessary expertise needed and are able also to represent and communicate with various segments of society will be important for effectiveness. Targeting the right key stakeholders, given collaborative missions and goals, allows access to broad arrays of social networks and resources and will engender trust in different segments of the community.

[24]See tc.securitypartnership.org/default.aspx?MenuItemID=101&MenuGroup=Home (accessed June 30, 2010).

Institutionalize Collaboration by Developing an Organizational and Operational Framework

Collaboration itself will be most effective if it is neutral—that is, nonpartisan, not-for-profit, and focused on providing benefit to the community at large (BENS, 2009). According to BENS, the legal, regulatory, and cultural barriers to collaboration often discourage long-term engagement by businesses when collaboration is government-funded and proscribed. The committee extends this observation to all organizations—neutral and nonpartisan collaboration is more conducive to trust building and creates an environment in which consensus can be built on common operating principles. The ideal organizing structure will reflect that neutrality—whether it is grounded in an existing community organization or incorporated as an independent 501(c)(3) organization—and will include the relationships necessary to coordinate preparedness efforts. A nonpartisan structure is less likely to exclude potential collaborators because of ideological differences, and is more likely to survive changes in political administration.

Collaboration organized by local governments can be effective—collaboration organized by the city of Seattle, Washington, being a notable example[25]—but the experience and observations of committee members leads the committee to conclude that relationships with the private sector are more easily formed and sustained when collaboration is not organized by a government agency, and that the organizational structure itself is likely to be more sustainable if not closely tied to a particular administration. Individual communities will need to decide which type of organizational structure would be most sustainable in their communities.

The organizational and leadership structures can be devised by using models from other communities and drawing on research and best practices or with technical assistance provided by a facilitator or nonprofit organization. Organizational aspects will vary by community, but it is important to provide for governance and ownership by local stakeholders.

BENS, MSU-CIP, and the LLIS series all recommend building collaboration, when possible, from the platform of an existing organization that has high credibility in the community (BENS, 2009; MSU, 2000; LLIS, 2006). For example, when invited by the governor of Iowa to explore feasibility of a partnership in Iowa, BENS went first to the Iowa Business Council, an organization that comprised the CEOs of the state's top 20 private employers, the presidents of three public universities, and the Iowa Bankers Association. The Iowa Business Council provided institutional endorsement and credibility for the partnership and aided its growth and expansion throughout the state. The Safeguard Iowa Partnership

[25]For example, the city of Seattle and King County, Washington, have formed a Vulnerable Populations Action Team (VPAT) that works with community-based organizations focusing on public health preparedness needs for their community members with special needs during times of disaster. See www.kingcounty.gov/healthservices/health/preparedness/VPAT/about.aspx (accessed September 15, 2010). Seattle has other private–public partnerships related to disaster preparedness and community disaster resilience. See www.cityofseattle.net/emergency/ (accessed September 15, 2010).

BOX 3.2
Illustrative Collaborative Model

Perhaps the best example of private–public collaboration that has moved to an advanced stage is the Safeguard Iowa Partnership (SIP). Originally facilitated by BENS, SIP was formally launched on January 29, 2007, with representatives of major Iowa businesses, the Iowa Business Council, and several state agencies. It is a voluntary coalition of Iowa's private and public-sector leaders, who share a commitment to strengthening the capacity of the state to prevent, prepare for, respond to, and recover from disasters. SIP partners work to reduce the impact of emergencies on their communities by pledging resources and offering support services.

SIP undertakes activities in five categories: resources and preparedness, communication and coordination, education and exercises, partnership development and outreach, and partnership marketing and public awareness. SIP's board of directors developed the five initiatives during a strategic planning session in 2008. The initiatives benefit SIP members, state government agencies, and the public.

SIP remains dedicated to increasing the participation of the private sector in its programs and therefore increasing the number and variety of assets available to lend to preparedness and response operations across Iowa. The partnership has developed a program to promote the establishment of organizational chapters within regions, counties, and cities. Chapters network between public and private-sector partners based in a given area with location-specific initiatives and information. SIP also pursues relationships actively with public-sector agencies, as evidenced by SIP's business seat in the State Emergency Operations Center (SEOC) and its involvement with the Iowa Department of Health and the Iowa Homeland Security and Emergency Management First Responders Advisory Committee.

SIP has been tested by disaster and its efficacy has been validated. During summer 2008, Iowa experienced a series of severe storms that produced several tornadoes and historic flooding. Over a four-week period, flood waters moved across Iowa and required the state to undertake extensive preparedness, response, and recovery operations. Overall, the 2008 summer storms resulted in 17 deaths, forced the evacuation of about 38,000 Iowans, and affected over 21,000 housing units.

During the 2008 summer storms, SIP helped to bridge the gap between Iowa's public and private sectors. SIP partners spent hundreds of hours during the 2008 summer storms contributing to Iowa's emergency response and recovery process, including assistance with general resource procurement at the SEOC.

SOURCE: www.LLIS.gov (accessed July 1, 2010).

was born as an initiative of the Iowa Business Council with BENS serving as a neutral facilitator.[26] It later incorporated as an independent 501(c)(3) organization. Stakeholders agreed on an operational framework that was institutionalized, and the Safeguard Iowa Partnership quickly grew—and was tested by disaster. (See Box 3.2 for a more detailed description of the partnership effort.)

[26]See www.safeguardiowa.org/ (accessed June 30, 2010).

The Seattle Project Impact effort was launched through the efforts of the Seattle Office of Emergency Management with the assistance of the Contingency Planners and Recovery Managers Group, consultants to the private and public sectors on issues related to emergency management preparedness and planning. Successful Seattle Project Impact programs were exported to surrounding jurisdictions and were all managed for years under the Seattle Project Impact operational framework. Seattle Project Impact supported the development of a separate but partnering nonprofit, the DRB Toolkit Workgroup,[27] who in turn partnered with Washington state communities to provide their tools to increase business disaster preparedness (Bullock et al., 2009).

Civic-minded organizations with executive-level volunteers are important in a partnership to provide both governance and operating support. Several partnership models suggest that collaboration can be governed and supported by multiple "teams": a high-level advisory council comprising CEOs and the directors of key state or local agencies to set strategic direction and an operating council that includes operations-level managers from business, civic organizations, and NGOs charged with program implementation. Sustaining commitment from a broad cross-section of members is critical for the success of a partnership.

Creating an organizational or governing structure with the conceptual model for resilience-focused private–public collaboration (Figure 2.1) in mind will help to ensure widespread acceptance as well as the efficacy and sustainability of a collaborative structure.

Identify Collective Resources and Capabilities that Mitigate Disaster Impact

As an early tool to build cohesiveness and a common sense of purpose, many organizations established to facilitate partnership development recommend that participants identify what their respective organizations can bring to their community in an emergency. The process is invariably an "eye opener" as it creates new understanding and trust among participants and lays the foundation from which to build new capability and resilience. Participants who recognize the availability of resources feel greater commitment to the process of collaboration when they recognize how sharing resources could benefit them. This inventory process can also provide early benefits by cataloging and coordinating identified resources in a systematic way. The Infrastructure Security Partnership[28] published a guide to building regional resilience that recommends a series of questions and steps that facilitate stronger resilience-focused collaboration among public and private stakeholders (TISP, 2006).

Collaboration can also serve as a means of forward thinking in the community. Through collaboration, for example, a community may develop a central community foundation to

[27]See www.drbtoolkit.org/ (accessed September 28, 2010).

[28]See www.tisp.org/ (accessed June 30, 2010).

serve as a repository for donated assistance funds for rapid distribution into the community when disaster strikes. Similarly, collaboration could result in initiatives that tie short-term benefits, such as improved bond ratings and community services, to actions that enhance longer-term preparedness and resilience.

Focus on Disaster Resilience, and Explore Community Resilience

Whether building collaboration from existing community organizations or beginning from scratch, one of the most important steps is to identify and agree on specific challenges, threats, or gaps in the community's disaster preparedness and resilience-building efforts that the new collaborative effort can address. It is important for those engaged in collaboration to share a commitment to the greater goal—the continuity of the community—as opposed to pursuing only parochial interests or self-interest. It is imperative to identify common issues related to emergency preparedness, for example, but it is also essential for collaborators to identify how emergency preparedness is part of a broader community-building effort. Such an effort was made in Arlington County, Virginia, following the attack on the Pentagon, located in that county, on September 11, 2001. The attack itself gave urgency to the need for resilience planning in the community, and community engagement followed because all sectors shared a similar vision for community resilience. A community most likely to survive disaster, according to Ron Carlee, Arlington County's manager until 2010, is one that actively commits to social equity and inclusion and creates a vision to which all its residents and institutions can relate (NRC, 2010b).

Develop Feasible and Measurable Objectives

Programmatically and financially sustainable collaboration depends on members' adoption of annual plans with well-defined, feasible, and measurable objectives; that exercise new capabilities; that deliver return on investment to all partners; and that manage growth and expectations. Examples of measurable annual program objectives include:

- The creation of a registry identifying private-sector resources and capabilities resident in the community—and points of contact for those resources—that could be mobilized in a disaster (the registry is a tangible product that increases local capacity and tangibly demonstrates the value of working together);
- The annual number of businesses and nongovernment organizations that participate in joint table-top or live exercises with government partners;
- A target number of private employers that use the collaborative partnership to strengthen disaster preparedness for their organizations and their employees;

- An annual increase—for example, by 10 percent—in the number of active participants and supporters; and
- Achieving financial and programmatic sustainability through a combination of public and private contributions and in-kind donations adequate to support at least one coordinator/staff.

Committee members have observed how many homeland security partnerships produce recommendations and plans and declare victory without delivering tangible results. Successful collaboration includes exercises to test and improve new capability. The results of these tests are tangible, measurable outcomes. Additionally, these actions enable capabilities to be perceived real assets, and exercising them on a continuing basis for many initiatives raises awareness, builds strong relationships, and prepares collaborators for any disaster. It is, however, difficult to know how some measures correlate with long-term benefits. Even so, allowing every collaborator and members of the community to perceive and measure value in collaboration provides incentives for continued participation. The challenges associated with choosing metrics are discussed in Chapter 4 and the research needs in this area are described in Chapter 5.

Build Capacity

An important role of disaster resilience-focused collaboration is to educate the community on community readiness. Effective capacity building will help ensure that critical services are available to the broader community during crises. Collaborative public-education initiatives and campaigns may include actions aimed at crisis mitigation, with end-result goals of building trust between local government and other support organizations, reducing risk, and shortening recovery time after extreme events. Capacity-building programs will need to include education and training about community resilience and its inextricable link with services provided by NGOs, FBOs, and other community organizations that often serve as the unofficial first responders to a disaster. Collaborative education efforts could assist organizations in establishing training programs for employees and members that increase the understanding of personal and organizational roles in disaster mitigation, preparation, response, and recovery.

Collaborate with Educational Institutions

Collaborating with local educational institutions increases access to local resources and capabilities. University scientists and technical experts may develop the fundamentals of a risk-education campaign on the basis of available research, elements of which can be tailored for elected officials, business leaders, and the broader community. Communication and education experts can be similarly tapped. Community colleges have many resources to offer especially given that 80 percent of the nation's first responders are credentialed at such institutions, according to the American Association of Community Colleges (AACC, 2006). Students in institutions of higher learning, including trade schools, can be enlisted to support resilience-building efforts and public outreach. At the same time, educational institutions can be encouraged to make business continuity and resilience education essential components of undergraduate education for economics and business majors, and to incorporate community resilience into the curricula of public policy and engineering disciplines. Collaborating with K-12 educational institutions can build on existing momentum for resilience-building activities within a community, for example, in the case of the Great California ShakeOut[29] drills. ShakeOut drills are being incorporated into school programs to fulfill annual earthquake drill requirements. Doing so could steer the next generation of leaders in all sectors to expect resilience-building to be a vital part of community economic, social, and environmental well-being. Partnering with K-12 educational institutions can help build capacity in a community's youngest members and their families.[30]

Rapid societal change and the resulting changes in community vulnerability suggest a need for comprehensive, continuing analysis, assessment, and research. The committee's conceptual model for resilience-focused collaboration (Figure 2.1) highlights the need for regular assessment of the community and of collaboration itself to ensure that goals and activities remain relevant. Although the committee understands that not every community will be able to do so, incorporating research directly into collaborative efforts will benefit collaboration and funders of collaboration by informing methods and metrics used. The assessment of the benefits of collaboration and of the direct and indirect costs of investing in collaboration could be better understood, and knowledge gained applied to other collaborative efforts. Decision making would be improved through direct input of research data. Incorporating random trial metrics in policy experiments by economists have shown some positive outcomes (e.g., Banerjee and Duflo, 2010; Banerjee et al., 2010). Positive outcomes have also been suggested through participatory research in the public health arena. The University of California, Berkeley, School of Public Health, and PolicyLink,[31] a national economic and social equity research and action institute, considered 10 case studies that reflect numerous public health issues in different locations. Their focus was on promoting public policy related to health through community-based participatory research. Studies included diesel bus pollution and its health consequences (Northern Manhattan, New

[29]See www.ShakeOut.org/ (accessed July 30, 2010).

[30]In California, a statewide earthquake preparedness exercise organized by the Earthquake County Alliance received the support of numerous county school superintendants who embraced the themes of the event and encouraged schools in their districts to engage students, families, local businesses, and community groups. Children were taught to secure their spaces in preparation for an earthquake, and taught how to be safe in the event of an earthquake. Resources were made available through schools for families and local organizations. See www.shakeout.org/schools/ (accessed August 24, 2010).

[31]See www.policylink.org/site/c.lkIXLbMNJrE/b.5136441/k.BD4A/Home.htm (accessed September 13, 2010).

York), environmental injustice in industrialized hog production (rural North Carolina), and lead exposure among children (Tar Creek, Oklahoma). The analysis highlighted sample policy and related outcomes that suggest the substantial role of partnerships and presented success factors and challenges faced across sites (Minkler et al., 2008).

Encourage Flexibility in Resource Administration

Whether support is given or received at the national or local levels, the ability to provide or use resources in a timely manner will be seriously hampered if too many conditions are tied to their use. Participants in the committee's workshop (NRC, 2010b) indicated that administering grants can be as time consuming as the activities they are intended to support. Some requirements were considered counterproductive. Requiring local matching funds as a condition of receiving resources, for example, can be prohibitive for rural or other communities in desperate need of support. It is essential to consider effective and flexible administration when providing grants and other funding support to allow creativity and the most effective use of resources.

It is also important that support is provided with the understanding that collaboration of the type described in this report needs long-term nurturing and may yield few short-term quantifiable outcomes. Funds provided without proper consideration of long-term benefits might actually create an environment of less productivity. Funding and resources provided for resilience-focused collaborative efforts will have greater impact if they provide incentives for groups to collaborate rather than encourage competition for limited funding. Funding mechanisms that encourage competition for grants, such as that incorporated by the DHS Urban Area Security Initiative (UASI), focus on short-term results and can be biased toward certain communities. The committee finds that such programs may actually create competition that is unproductive in the long term in order to realize short-lived benefits. Further, funds directed to specific communities or outcomes may ignore the greater good done through collaboration elsewhere. More inclusive funding programs that are less targeted to specific agencies or outcomes may be more beneficial to communities in the long term.

CREATING AN ENVIRONMENT FOR CHANGE

Community resilience is more than the ability to conduct disaster response, and private–public sector collaboration is an optimal means of generating community resilience. In preparing this report, the committee faced a daunting challenge: to identify specific aspects of private–public sector collaboration most crucial for building community disaster resilience in a broader context. Box 3.3 provides a concise and overarching summary of the guidelines provided in this report, offering guidance on how the sociopolitical environment might

BOX 3.3
Overarching Guidelines

The committee was tasked with developing a set of guidelines for private-sector engagement in enhancing community disaster resilience, but finds that its overarching guidelines are applicable to all sectors. The guidelines were designed to address community-level private–public collaboration for enhancing disaster resilience, but they will also apply to collaboration—or those wishing to support collaboration—at any level. These guidelines can be used in concert with the committee's conceptual model for resilience-focused private–public sector collaboration (Figure 2.1), which shows the relationship between collaborative elements and outcomes. Keeping in mind how different elements of collaboration are related may facilitate more successful application of the guidelines.

1. Pursue community-level private–public sector collaboration as a fundamental component of community resilience in general and disaster resilience in particular. Resilience-focused private–public collaboration ideally will:

    a. Integrate with broader capacity-building efforts within the community and include all community actors.
    b. Emphasize principles of comprehensive emergency management allowing preparation for all hazards and all phases of the disaster cycle to drive goals and activities.
    c. Function as a system of horizontal networks at the community level, coordinating with higher government and organizational levels.
    d. Develop flexible, evolving entities and establish processes to set goals, conduct continuing self-assessment, meet new challenges, and ensure sustainability.
    e. Institutionalize as a neutral, nonpartisan entity with dedicated staff.

2. Build capacity through communication and training programs for those engaged in private–public collaboration and for the broader community. Resilience-focused private–public collaboration ideally will:

    a. Incorporate capacity building into collaboration from the onset.
    b. Target educational campaigns toward crisis mitigation with goals of community readiness, continuity planning, trust building, risk reduction, and shortened recovery time.
    c. Encourage all organizations in the private and public sectors to commit to organizational resilience through business-continuity measures.
    d. Partner with educational institutions in developing educational campaigns and disseminating information.
    e. Institutionalize the practice of embedding research into resilience-focused private–public sector collaboration by building research directly into existing and future collaborative efforts.

3. Respect well-informed, locally determined all-hazards preparedness and resilience priorities.

4. Develop funding and resource allocation strategies that are flexible in administration.

foster community-level partnership building more effectively. Although the guidelines presented address resilience-focused private–public collaboration at the community level, they are applicable to collaboration at any level.

The private sector can build capacity, for example, by educating local elected officials about the benefits of participation in and support of community cross-sector partnerships and collaboration that encourage anticipatory risk reduction. It can combine the power of for-profit and nonprofit organizations to influence legislation and policy that support resilience-focused disaster mitigation and business continuity planning at the local, state, and federal levels. At the same time, the private sector, including NGOs and FBOs, can commit to internal organizational resilience through business-continuity measures, and encourage preparedness for employees and their families through education and training, activities, and incentives.

The public sector is typically regarded as a leader in providing disaster response and recovery aid. It is essential, then, that the public sector promote activities that increase knowledge about resilience, resilience building, and the importance of private–public collaboration among community members. Government employees may be trained to promote resilience in their own lives and to understand their roles in the continuity of their organizations during and following a disaster.

Federal partners, like community-level counterparts, could learn from unsuccessful efforts to develop strategies for mainstreaming collaboration in existing programs. Training and learning experiences aimed at developing the skill necessary for forming, sustaining, and institutionalizing private–public collaboration could be built on such lessons learned. Federal activities could include producing training materials for disaster personnel, placing a greater emphasis on partnership-building skills in programs offered by the Emergency Management Institute, funding workshops and train-the-trainer experiences, sponsoring the development of higher-education courses and textbooks on the topic, and providing learning experiences for members of the federal workforce.

REFERENCES

AACC (American Association of Community Colleges). 2006. First Responders: Community Colleges on the Front Line of Security. Washington, DC. Available at www.aacc.nche.edu/Publications/Reports/Documents/firstresponders.pdf (accessed September 16, 2010).

Alesch, D., P. May, R. Olshansky, W. Petak, and K. Tierney. 2004. *Promoting Seismic Safety: Guidance for Advocates*. MCEER-04-SP02. Prepared for Federal Emergency Management Agency, Washington, DC. Buffalo, NY: The State University of New York at Buffalo.

Banerjee, A. V. and E. Duflo. 2010. Giving Credit Where it is Due. *Journal of Economic Perspectives*. Paper available at econ-www.mit.edu/files/5416 (accessed August 4, 2010).

Banerjee, A. V., E. Duflo, R. Glennerster, D. Kothari. 2010. Improving Immunization Coverage in Rural India: A Clustered Randomized Controlled Evaluation of Immunization Campaigns with and without Incentives. *British Medical Journal* 340: c2220.

BENS (Business Executives for National Security). 2009. Building a Resilient America: A proposal to strengthen private–public collaboration. March 3. Available at www.bens.org/PBO Proposal_03_04_09.pdf (accessed March 12, 2010).

Bevc, C. 2010. Working on the Edge: Examining the Dynamics of Space-Time Covariates in the Multi-Organizational Networks Following the September 11th Attacks on the World Trade Center. Doctoral dissertation, Dept. of Sociology, University of Colorado at Boulder.

Briggs, R. O., G. Kolfschoten, C. Albrecht, D. R. Dean, and S. Lukosch. 2009. A Seven-Layer Model of Collaboration: Separation of Concerns for Designers of Collaboration Systems. *Proceedings of the International Conference on Information Systems*. Association for Information Systems. Available at aisel.aisnet.org/cgi/viewcontent.cgi?article=1179&context =icis2009 (accessed July 1, 2010).

Bullock, J. A., G. D. Haddow, and K. S. Haddow (editors). 2009. *Global Warming, Natural Hazards, and Emergency Management.* Boca Raton, FL: CRC Press.

CDC-ATSDR (Center for Disease Control-Agency for Toxic Substances and Disease Registry). 1997. Principles of Community Engagement. Atlanta, GA: Center for Disease Control and Prevention. Available at www.cdc.gov/phppo/pce/ (accessed July 1, 2010).

FEMA (Federal Emergency Management Agency). 2008. National Response Framework. Washington, DC: U.S. Department of Homeland Security. Available at www.fema.gov/pdf/emergency/nrf/nrf-core.pdf (accessed March 11, 2010).

Fleischman, A. R. 2007. Community engagement in urban health research. *Journal of Urban Health* 84(4): 469-471.

Kadlec, A., and W. Friedman. 2008. Public Engagement: A Primer from Public Agenda. *Essentials* No. 01/2008. New York: Center for Advances in Public Engagement. Available at www.publicagenda.org/files/pdf/public_engagement_primer_0.pdf (accessed July 1, 2010).

Kapucu, M. 2007. Non-profit response to catastrophic disasters. *Disaster Prevention and Management.* 16: 551-561.

Lasker, R. D., E. S. Weiss, Q. E. Baker, A. K. Collier, B. A. Israel, A. Plough, and C. Bruner. 2003. *Journal of Urban Health* 80(1): 14-60.

Leighninger, M. 2009. The Promise and Challenge of Neighborhood Democracy: Lessons from the intersection of government and community. A report on the "Democratic Governance at the Neighborhood Level" meeting, November 11, 2008, Orlando, FL.

LLIS (Lessons Learned Information Sharing). 2006. Public-Private Partnerships for Emergency Preparedness. LLIS.gov Best Practice Series. Available at oja.wi.gov/docview.asp?docid=14758&locid=97 (accessed July 1, 2010).

Lukensmeyer, C. J., and L. H. Torres. 2006. Public Deliberation: A Manager's Guide to Citizen Engagement. *Collaboration Series.* Washington, DC: The IBM Center for the Business of Government. Available at www.businessofgovernment.org/sites/default/files/LukensmeyerReport.pdf (accessed August 31, 2010).

Magsino, S. 2009. *Applications of Social Network Analysis for Building Community Disaster Resilience: Workshop Summary.* Washington, DC: The National Academies Press.

Mileti, D. S., 1999. *Disasters by Design: A Reassessment of Natural Hazards in the United States.* Washington, DC: The Joseph Henry Press.

Milward, H. B. and Provan, K. G. 2006. A Manager's Guide to Choosing and Using Collaborative Networks. *Networks and Partnerships Series.* Washington, DC: The IBM Center for the Business of Government. Available at www.businessofgovernment.org/sites/default/files/CollaborativeNetworks.pdf (accessed September 2, 2010).

Minkler, M., V. B. Vásquez, C. Chang, J. Miller, V. Rubin, A. G. Blackwell, M. Thompson, R. Flournoy, and J. Bell. 2008. Promoting healthy public policy through community-based participatory research: Ten case studies. A project of the University of California, Berkeley, School of Public Health and PolicyLink, funded by a grant from W. K. Kellogg Foundation. Available at www.policylink.org/atf/cf/%7B97C6D565-BB43-406D-A6D5-ECA3BBF35AF0%7D/CBPR_PromotingHealthyPublicPolicy_final.pdf (accessed September 10, 2010).

Morris, A. 1984. *The Origins of the Civil Rights Movement.* New York: The Free Press.

MSU (Michigan State University). 2000. Critical Incident Protocol—A Public and Private Partnership. Project supported by Grant No. 98-LF-CX-0007 awarded by the U.S. Department of Justice. Available at www.cip.msu.edu/cip.pdf (accessed July 1, 2010).

NRC (National Research Council). 2006. *Facing Hazards and Disasters: Understanding Human Dimensions.* Washington, DC: The National Academies Press.

NRC (National Research Council). 2010a. *Adapting to the Impacts of Climate Change.* Washington, DC: The National Academies Press.

NRC (National Research Council). 2010b. *Private–Public Sector Collaboration to Enhance Community Disaster Resilience: A Workshop Report.* Washington, DC: The National Academies Press.

Patterson, O., F. Weil, and K. Patel. 2010. The role of community in disaster response: Conceptual models. *Population Research and Policy Review* 29(2): 127-141.

TISP (The Infrastructure Security Partnership). 2006. Regional Disaster Resilience: A Guide for Developing an Action Plan. Reston, VA: American Society of Civil Engineers.

Wenger, E. 1998. *Community of Practice: Learning, Meaning and Identity.* Cambridge, UK: Cambridge University Press.

CHAPTER FOUR

Challenges to Sustainable Resilience-Focused Collaboration

Achieving and sustaining community resilience are in the interest of the nation, states, communities, businesses, and citizens. Why, then, do resilient communities seem the exception rather than the rule? A partial answer to that question lies in the wide array of challenges that inhibit or block efforts to create the collaborative context required to achieve community resilience. The committee acknowledges the growing attention in this country to community disaster resilience in general and to resilience-focused private–public collaboration specifically. It also acknowledges that although numerous individual programs provide support for specific efforts, a political and social environment truly supportive of the development of community-based, sustainable, resilience-focused private–public collaboration does not exist at the national level. That, in a sense, leaves communities to determine independently how to move forward, what works, what is sustainable, and—often by trial and error—what does not work. Resources or incentive to start again following failed efforts may not exist. It is, however, in the best interest of communities to make the effort. Private–public collaboration to enhance resilience can be extremely effective when efforts are designed to be largely autonomous at the community level with ties to higher levels of government for additional support and expertise.

As a community moves forward to adopt and apply a framework for collaboration, whether the one provided in this report or another, sensitivity to the inevitable challenges is necessary. Some issues that may impede successful and sustainable private–public collaboration are described below. They have been identified by committee members and by participants in the committee's information-gathering workshop (NRC, 2010). Some of the challenges described here may be in the category of wicked problems (discussed in Chapter 2). Some are encountered at many levels of government, and indeed, the committee offers examples that would be most familiar to the Department of Homeland Security (DHS), the sponsor of the present study, and that provide lessons that are scalable down to the community level. Recommended research to address some of the challenges described is discussed in Chapter 5.

INCREASING CAPACITY AND ACCESS OF THE VULNERABLE

The most vulnerable in a community often do not have the capability, capacity, or access needed to participate in resilience-focused private–public collaborative efforts. The U.S. population is diverse, and the capacity to adopt resilience-enhancing measures, including forming or participating in private–public sector collaboration, varies considerably. A major factor preventing more widespread capacity development is that some groups are highly vulnerable, at risk for extreme events, and routinely subject to economic and social stressors. Disaster preparedness and resilience are not often on the agendas of those who deal regularly with chronic conditions and crises, such as poverty, crime, violence, serious illness, and unemployment. In addition, many groups in the United States lack firm connections to mainstream community institutions that could serve as sources of disaster-related information and social support. Such groups include non-English-speakers, people who have mental health and substance-abuse problems, elderly single persons living alone (a growing segment of the population), people who have physical disabilities, those who are homeless, and those who live in communities on a transient basis. That is not to argue that such groups lack organization and social solidarity (although many people in U.S. communities do suffer from social isolation). But the people in those groups and the organizations that serve them may not have the knowledge and access to information that would motivate or allow them to engage in resilience-enhancing collaborative efforts.

The aftermath of Hurricane Katrina is a vivid example of how the poor, minority, elderly, and infirm groups have not been well served in response or recovery efforts (Colten et al., 2008). Planning for evacuation in the face of an impending hurricane was extensive throughout the Gulf of Mexico region, and the evacuation before Katrina was considered largely successful. However, the needs of those who were dependent on public transportation were not taken into account (Townsend, 2006). In the days after Katrina, those left in New Orleans—including institutionalized populations and those who served them—were forced to endure extreme hardship and in many cases lost their lives. The Katrina example is not unique. The ways in which social inequality and diversity affect the ability to absorb and recover from the effects of disaster have been well documented, and social vulnerability itself is a major subject of study in disaster research (e.g., Tierney, 2007; NRC, 2006; Cutter et al., 2008).

As stated several times in this report, successful resilience-building through private–public collaboration depends on the inclusion of the full fabric of the community. Community resilience will be improved only if strategies that identify and engage the vulnerable populations in the community and in the organizations that represent them are considered and used. Addressing vulnerabilities reduces the need for response and recovery. Failing to identify vulnerable segments of the population leaves the entire community less resilient when disaster occurs.

PERCEPTIONS OF RISK AND UNCERTAINTY

Individuals, institutions, and entire sectors often do not perceive that hazards pose unacceptable risk or that they may have a responsibility—or even the capacity—to reduce the risk. Successful resilience-focused private–public collaboration depends partially on increasing the transparency and general knowledge of risk and uncertainty. Successful collaborative strategy-building attempts to account for the lack of understanding among community members of what constitutes an extreme event and the perception that an extreme event will not affect an individual personally.

Perception is the basis of action, and inaccurate perceptions stand in the way of concerted action to promote community disaster resilience. Individuals, groups, and societies have great difficulty in understanding and acting on information related to low-probability–high-consequence events. Understanding risk is conceptually difficult and subject to biases, including focused attention on a recent or dramatic event (often to the exclusion of more probable events) or expectations about future events based on past events. As an example of the latter bias, in the case of Hurricane Katrina, evidence indicates some New Orleans minority residents chose not to evacuate their homes in spite of a mandatory evacuation order because of past experience with Hurricanes Betsy and Camille (Elder et al., 2007). They reasoned that because they were safe at home during previous storms, there would be little danger for them from Katrina; how much worse could Katrina be?

Time horizons also affect perception of risk. People may believe that a major disaster is likely to occur but not in their own lifetimes. And individuals and institutions may have the tendency to think and plan in terms of relatively short periods. That may be part of the reason why political leaders discount the future benefits of making their communities more resilient to rare events, especially if their terms in office are relatively short. Without a motivating sense of urgency, the benefits of participating in collaborative efforts may not be appreciated. Even if there is a general sense of the likelihood of a particular type of disaster, such as Californians' wide recognition of the likelihood of earthquakes, people may find it difficult to believe that such an event will affect them personally.

Another challenge to be considered in developing collaborative strategies is related to people's general inability to grasp the concept of uncertainty. Predictions about the future—including likelihood of disasters—always contain elements of uncertainty. However, when uncertainties appear unacceptably large, people will not act or will postpone taking action. For example, two key elements in the inability of the public and of institutions to appreciate and act on climate change are the uncertainties surrounding projections of the effects of climate change and the uncertainties related to projections on meaningful geographic and time scales (NRC, 2009). The same is true for other types of hazards: when the perceived uncertainty associated with an event and its consequences is high, action is difficult to justify.

An important caveat is that judgments concerning "acceptable" levels of uncertainty, like those concerning acceptable risk, are social rather than scientific.

Research on societal responses to natural, technologic, and other threats shows that there is no necessary connection between understanding risks and taking action on the basis of understanding (NRC, 2006). As discussed above, even when well aware of the risks faced, some groups lack the capacity to take recommended measures, for example, because of financial, health, mental health, and language issues. The large literature on factors that affect disaster preparedness shows that better-off segments of the public—as measured by income, education, and home ownership—are generally better prepared than their less well-off counterparts (NRC, 2006).

Successful resilience-focused collaboration includes strategies to encourage organizations to develop established processes for recognizing threats and evaluating risk. That encouragement is a fight against cultural momentum. Businesses and other private-sector organizations are influenced by inaccurate and incomplete perceptions of risk and therefore might not provide the resources to mitigate risks or recognize the potential value of collaboration. The concept of enterprise risk management (ERM) has taken hold to some degree in the private sector, where it helps some firms assess their risks on an organization-wide basis, set priorities among risks, and develop consistent, comprehensive approaches to risk management.[1] However, ERM is not widely practiced in the private sector and is even less prevalent in the public sector.

SCALES OF COLLABORATION

Local, regional, and national collaborative efforts are not effectively linked or harmonized. That can present a challenge to people engaged in community-based private–public collaboration as they try to identify and leverage community resources and plan implementation strategies. A mismatch exists between the scales on which many organizations operate and the scales on which resilience-enhancing actions need to be taken, sometimes making it difficult to sustain collaboration like that described in this report. Some businesses and nongovernment organizations collaborate with DHS at the national level but do not participate in local collaborative efforts in the communities where they have a physical presence. It may be difficult for a large corporation such as a national-scale retail chain to engage locally with the full fabric of the community on an ongoing basis, to collaborate nationally with DHS and Federal Emergency Management Agency (FEMA) policy makers and planners and to coordinate with other businesses on crisis supply-chain issues. Other businesses may be very active locally but are not part of regional or national collaborative

[1]For example, see the Casualty Actuarial Society Web site (www.casact.org/research/erm/; accessed June 18, 2010) for more information on ERM.

efforts coordinated by DHS and others who can provide the strategic context and programmatic funding opportunities for local efforts.

On the public-sector side, the DHS regional and local presence is fragmented, partial, and still evolving, so it is difficult to be aware of or enable local and regional resilience-focused collaboration and thus difficult for community-level collaborative structures to network vertically with them. Although some agencies in DHS (such as the Coast Guard, Customs and Border Patrol, and the Transportation Security Agency) do have a local presence in some parts of the country and some sectors, their ties to local-level private entities are generally mission-specific rather than focused more broadly on enhancing community resilience to all hazards. FEMA has influenced and will continue to influence local resilience-building actions, for example, through its responsibilities under the Disaster Mitigation Act of 2000 and the Stafford Act, but it has no physical presence below the regional level. It is then a challenging task for communities to identify which vertical networking points in the private and public sectors are vital and to plan their strategies accordingly.

DIVERGING INTERESTS

The interests of collaborators often diverge, and this impedes the development of trusted collaborative relationships. When diverse stakeholders engage in a joint venture, vested interests often come into play and can result in conflict and failure to agree on objectives, goals, and methods. No entity can be faulted for pursuing its own interests; doing so is natural and understandable. Problems develop, however, when actors view collaboration as a zero-sum game. Such problems can complicate resilience-enhancing efforts and the development of effective collaboration. Organizations want assurances that the benefits of engaging in collaboration outweigh the perceived loss of autonomy, the financial and reputation-related risks, and the costs associated with investment in collaborative activities. When there is a failure to provide tangible and meaningful rewards to participants in collaboration, problems develop. However, it is also important to build confidence among collaborators that working for the broader collective good benefits the individual collaborator. The building of community and societal resilience depends on the ability to acknowledge and address the priorities of diverse parties while defining and leveraging common interests through collaborative effort.

The DHS Voluntary Private Sector Preparedness Accreditation and Certification Program (PS-Prep) is an example of how public- and private-sector interests diverge.[2] DHS

[2]See www.fema.gov/privatesector/preparedness/index.htm (accessed June 18, 2010). The impetus for the PS-Prep program was Title IX of the Implementing Recommendations of the 9/11 Commission Act of 2007 (Public Law 110-53, 2007), which sought to increase resilience by providing incentives for private-sector preparedness. In January 2009, DHS began a series of public stakeholder meetings on the topic of standards. In October 2009, three standards for private-sector preparedness were recommended by DHS as part of PS-Prep: NFPA 1600, ASIS International SPC.1-2009, and British Standard 25999 (NFPA, 2007; ASIS International, 2009; BSI Group, 2009). A period of public comment regarding the standards followed.

established PS-Prep as a top-down effort to promulgate voluntary resilience standards for businesses, by which collaboration is conducted in a formal process centrally managed by DHS. Predictably, private-sector responses have been mixed. The business-continuity community, a private-sector group that stands to gain considerably from the existence of the program, has been active in disseminating information about the new voluntary standards. A commentator from that sector expressed concern that the program was not receiving the emphasis warranted within DHS but also noted that critics of PS-Prep see it as "a back-door way of DHS to regulate industry and impose additional rules, regulations, and costs upon the private sector."[3] A May 2009 blog posting by IBS Publishing featured an interview with a San Francisco-area banking executive quoted as saying that "banks could endure a compliance nightmare" as a consequence of the new standards and that "while the idea is excellent, it threatens the banking industry. . . . So how do we—the public sector and the private sector—play together?"[4] A cursory look at comments raised in public meetings reveals private-sector concerns regarding the economic burden and the training required for personnel to ensure that businesses are in compliance. Small businesses especially would feel the burden. In summary, PS-Prep has had the mixed result of defining widely accepted standards and metrics but through a process that became a deterrent to local collaborative efforts. The lesson learned from PS-Prep is scalable to the community level: it is essential for communities establishing private–public collaboration to be sensitive to and identify collaborators' sometimes competing self-interests. It is necessary to identify incentives that will engage all sectors of the population for the community to embrace the goals and methods of collaboration.

It is best to avoid conflict and competition among those engaged in private–public collaboration and in the community more broadly. Another national-level example of how vested interests can create conflict and competition is the DHS Urban Areas Security Initiative (UASI),[5] which was intended to increase community and regional preparedness against terrorist attacks and other extreme events. From its inception, the initiative was focused more on traditional crisis-relevant organizations, such as fire and police departments and local emergency-management agencies, and less on other types of organizations, such as businesses, public health agencies, school districts, community-based organizations, and universities. The program was marked by various types of competition and conflict, not only among agencies at the community level but between core cities in regions and their less urbanized counterparts. Even communities receiving UASI grants saw themselves as vying against one another for funding. Competition was created simultaneously with collaboration at the community level as different community agencies sought funding on the basis of their own definitions of what was needed to combat terrorism while attempting to

[3]See securitydebrief.adfero.com/2009/11/03/private-sector-prep-does-anybody-care/ (accessed August 4, 2010).
[4]Available at www.zoominfo.com/people/Cardoza_Barry_312319040.aspx (accessed June 30, 2010).
[5]See www.fema.gov/government/grant/uasi/index.shtm (access June 18, 2010).

stay within program guidelines. DHS used a centralized hierarchic approach to program development and funding for management-control and accountability reasons, and this resulted in a program that has not effectively supported and enabled community-level collaboration.

There is always a political dimension in an effort to form a collaboration of citizens, community organizations, and businesses. The community organization process itself can be coupled with and supported by a political agenda or can be seen as a threat by political parties or candidates. Once empowered, private–public collaboration may challenge existing assumptions that are themselves embedded in politics, such as assumptions about the need for unfettered community growth even in areas vulnerable to floods or other locally known threats and even if such growth may lead to larger disaster losses. To the extent that community-based coalitions become involved in debates over land use and codes, government priorities, taxes, government accountability, provision of assistance to groups to enable them to become more resilient, and participation in federal programs, their activities will be framed as political and responded to accordingly.

TRUST AMONG COLLABORATORS

Overall, there is a lack of trust among parties that collaborate to build resilience. Federal agencies compete over program dollars. State and local agencies resent federal interference. Businesses fear government regulation, direction, or control that will limit creativity and market flexibility. There is a wide cultural gap between private-sector managers and public-sector officials. Their organizational cultures, standards, and languages are different. There are too few opportunities and minimal motivation to build relationships and trust. Building the trusting relationships necessary for collaboration requires the mutual understanding of the motivations and needs of stakeholders. Once trust and collaborative relationships have been developed, there is a need to nurture them constantly. Sustainability of collaboration is dependent on collaborators trusting that the collaborative structure and strategies are correct, on their familiarity with the strengths and resources of the collaborative network, and on their commitment to collaboration for the long haul.

In a Council on Foreign Relations (CFR) white paper on private–public partnerships, Flynn and Prieto (2006) traced barriers to DHS participation in collaboration and in effective enabling of the development of a community-based culture of sustainable resilience. For example, DHS management was made up primarily of personnel from the agencies that merged to form DHS and of temporary detailees from other agencies, many of whom may have remained more loyal to their parent agencies than to DHS. The agency relies heavily on contractors, including contractors in such key fields as policy and strategy development. Turnover tends to be high, morale tends to be comparatively low, and DHS personnel generally lack familiarity with the needs and resources of the private and nonprofit sectors. Those circumstances make it difficult for DHS to enable or participate effectively in private–public partnerships, and the same types of circumstances will pose problems in building resilience at the community level. The CFR report points to the need to "strengthen the quality and experience of DHS and establish a personnel exchange program with the private sector to help make DHS a more effective partner to the private sector" (Flynn and Prieto, 2006:35). It is also important for community-level collaboration to consider how to familiarize those engaged with the needs and resources of other collaborators and how to build trust among them. There are examples of effective local and regional collaboration led by DHS agencies that could be used as models. For example, the U.S. Coast Guard supports local private–public harbor-safety committees and regional area-security committees that bring together government, private, and nonprofit users of ports and waterways to collaborate on safety and security issues. The Coast Guard and the National Research Council's Transportation Research Board co-sponsor an annual conference for those committees.[6]

INFORMATION SHARING

Incomplete and ineffective sharing of information concerning threats and vulnerabilities constitutes a challenge to private–public collaboration. Both government and the private sector have legitimate concerns regarding the sharing of information. The private sector's concerns include the sensitivity of its information, legal limits on information disclosure, advantages that competitors might gain through sharing, and the existence of business-to-business contracts, such as nondisclosure agreements. Private–public information sharing is often perceived as lacking appropriate balance: regulations require businesses to disclose information to government, but government may not reciprocate with information that businesses need (Flynn and Prieto, 2006).

Government agencies are also subject to privacy restrictions, transparency requirements, and security rules. They are required to protect classified information and information considered "sensitive but unclassified" and "for official use only." At the same time, lower-level government entities and entities outside government may require such information for their own preparedness activities but must have security clearances. Those holding the information decide which entities should receive such clearances and how extensive the information dissemination should be. If key data are withheld from communities, it is conceivable that rigorous analysis of infrastructure vulnerabilities may not be possible. This may create doubt about the effectiveness of resilience-focused collaborative efforts among those engaged and the community that could lead to mistrust. Assessment of community vulnerabilities and resources is an early step of collaboration forming suggested by the committee.

[6] See www.trb.org/marinetransportation1/calendar1.aspx (accessed June 20, 2010).

Concerns about terrorism have only increased those tensions, making even government-to-government information sharing difficult. Local communities have long contended that they should have access to threat-related information that can support their risk-management and emergency-management decision making, but collaboration between government offices on such issues has proved problematic. Federal officials have been reluctant to share threat information with local first responders and with elected and appointed officials. Further concerns are raised when law-enforcement agencies do gain access to threat information but local government leaders do not. For example, in 2005, the mayor of Portland, Oregon, ended the city's participation in the multiagency Joint Terrorism Task Force because he was denied access to information that had been provided to Portland's own law-enforcement officials. Such cases highlight the challenge of establishing a balance between the need to keep sensitive information out of the hands of terrorists and the need to support those responsible for protecting the public in the event of a terrorist attack (for more discussion, see Flynn and Prieto, 2006; GAO, 2005, 2008).

Several participants of the committee's information-gathering workshop indicated that a private sector actor may hesitate to participate in private-public sector collaboration if it perceives that data or responsibilities are not shared equitably at the community level (NRC, 2010). The perception that one party, organization, or sector may carry a greater burden than another—and perhaps a greater liability because of that burden—can deter collaboration. The challenge of balancing the needs to share and to keep information by the various sectors at the community level will have to be carefully addressed by those involved in private–public collaboration.

Those collaborating might consider information as a resource and understand the limitations in the availability (and accuracy) of information as strategies are developed, activities are coordinated, and responses are implemented.

SPANNING BOUNDARIES

Organizations often do not seek, develop, or reward the organizational and individual competences needed to support collaborative efforts. Particular types of individual skills and expertise and particular types of organizational entities are required to build trusted relationships and foster collaborative action that overcomes interorganization and intergovernment boundaries. Collaboration and partnerships are often forged through the efforts of "boundary-spanning" people who venture outside their organizational cultures and are open to views and concerns of other organizations. Organizations wishing to build collaborative relationships may find they must assign boundary-spanning responsibilities to appropriate people and empower them to act. Often, however, that important role is neglected, and boundary-spanning activities are constrained. For example, public-sector officials are trained in how *not* to build relationships with private-sector managers (refusing free meals and setting contracting restrictions and privacy requirements) instead of how to foster them. When public-sector entities interact with private ones, the interactions often center on legal and regulatory issues as opposed to voluntary and mutually beneficial collaboration. As noted above, information-sharing restrictions hamper collaboration both within and between the public and private sectors.

Effective collaboration is based on mutual understanding. However, personnel in the public, for-profit, and nonprofit sectors tend to lack cross-sectoral understanding and the capacity to obtain it. The typical business school curriculum contains little or no content on public-sector management, particularly risk and emergency management. Career civil servants may have minimal experience with and knowledge of the operations of private businesses. Similarly, both the public and for-profit sectors lack an understanding of the challenges associated with nonprofit management.

Communities benefit when they grasp the lack of common understanding and framework in fields associated with disaster risk and resilience; they also benefit from learning about examples of boundary-crossing success. For example, because of its mission in critical-infrastructure protection and because the vast bulk of that infrastructure is in private hands, DHS has the opportunity to interact with utility service providers, the banking and financial sector, as well as the other federally designated critical infrastructure sectors. FEMA administers the National Flood Insurance Program,[7] which requires the agency to have relationships with private insurers and reinsurers. FEMA also interacts with the private sector on key loss-reduction issues, such as building-code provisions. The Citizen Corps program works directly with members of the public in an effort to strengthen civil-society disaster-response capabilities.[8] Such entities as the Institute of Business and Home Safety,[9] an insurance group, and Business Executives for National Security (BENS)[10] have ties with federal government agencies. As noted earlier in this report, private–public partnerships—such as the American Lifelines Alliance, a partnership between FEMA and the American Society of Civil Engineers—provide occasions for interactions focusing on infrastructure resilience.[11]

Organizations, like individuals, may span boundaries. Science and technology literature and literature from such fields as environmental and climate-change policy emphasize the role of "boundary organizations" in creating necessary linkages and exchanges among entities and sectors that would otherwise not be able to understand or work well with one another. The boundary organization concept was originally developed from research on interactions among science and policy communities (see Guston, 1999, 2000, 2001), but it

[7]See www.fema.gov/business/nfip/ (accessed June 23, 2010).
[8]See www.citizencorps.gov/ (accessed June 23, 2010).
[9]See www.disastersafety.org/ (accessed June 23, 2010).
[10]See www.bens.org/home.html (accessed June 23, 2010).
[11]See www.americanlifelinesalliance.org/ (accessed June 23, 2010).

has also been used in discussions on such topics as climate-related decision support (NRC, 2009). In that domain, for example, boundary organizations are seen as playing a useful and productive role in supporting interactions between scientists and users of scientific information. Such organizations facilitate communication not only between scientists and other constituencies, but also among diverse stakeholders; they help sustain interaction over time and provide an environment in which interorganizational conflict and competition can be minimized. The National Institute of Building Sciences (NIBS) and its Multihazard Mitigation Council (MMC; see Box 4.1), the Applied Technology Council (see Box 4.2), and the Natural Hazards Center (see Box 4.3) are examples of boundary organizations that seek to address specific types of disaster-resilience needs. They perform many useful functions, as indicated by their longevity and ability to attract resources, the use of their products, and their influence on loss-reduction policies and practices (MMC, 2005).[12] However, the niches that boundary organizations occupy are relatively narrow compared with the holistic goals of all-hazards resilience enhancement. It is equally important that none of those or the many other boundary organizations are explicitly concerned with working with diverse constituencies on broad-based resilience activities.

To create productive private–public collaboration, more time and effort will need to be devoted to facilitating multisector collaboration for enhanced disaster resilience. Providing training and educational experiences for private- and public-sector personnel, offering incentives to those who engage in boundary-spanning activities, and supporting and expanding the activities of boundary organizations whose missions are consistent with resilience goals will bring progress toward that goal. Multisector collaboration is unlikely on broad scales unless action is also taken at the national level to address the fragmentation and lack of coordination, discussed below, that currently characterize societal efforts to improve disaster resilience. If the status quo is allowed to persist, resilience-focused collaboration will continue to be narrow, specialized, noninclusive, uneven, and uncoordinated across all sectors of society and in terms of resilience objectives.

FRAGMENTATION, INCONSISTENCIES, AND LACK OF COORDINATION

Although the United States had previously embraced a comprehensive, all-hazards approach to emergency management, the events of September 11, 2001 led to, among other things, a host of new programs and funding opportunities to enhance community resilience. It also, however, helped to create a bifurcated national emergency-management system that both elevated terrorist threats above other threats and led to the proliferation of separate systems for terrorism and hazards management at state and local levels. Executive orders,

[12]For example, the NIBS MMC study was conducted in response to a congressional mandate and demonstrated through the use of rigorous analytic approaches that investments in mitigation result in savings to the nation and the federal treasury.

BOX 4.1
The National Institute of Building Sciences

The National Institute of Building Sciences (NIBS) and its Multihazard Mitigation Council are examples of boundary organizations that seek to address specific types of disaster-resilience needs. According to their Web site, NIBS

> . . . is a non-profit, non-governmental organization that successfully brings together representatives of government, the professions, industry, labor and consumer interests, and regulatory agencies to focus on the identification and resolution of problems and potential problems that hamper the construction of safe, affordable structures for housing, commerce and industry throughout the United States. Authorized by the U.S. Congress, the Institute provides an authoritative source and a unique opportunity for free and candid discussion among private and public sectors within the built environment.[a]

The Multihazard Mitigation Council (MMC), one of several councils that operate under the auspices of NIBS, was responsible for the report *Natural Hazard Mitigation Saves* (MMC, 2005), which established that investments in mitigation projects and process activities are cost-effective. Like other NIBS councils, the MMC provides a continuing venue for discussion among entities in the private and public sectors. MMC members include universities, state officials, a federal agency (the National Institute of Standards and Technology), professional societies, producers of safety devices, and engineering consulting firms.

[a] See www.nibs.org (accessed June 23, 2010).

BOX 4.2
The Applied Technology Council

The Applied Technology Council (ATC), a nonprofit organization in Redwood City, California, is another type of boundary organization. Founded in 1973 by members of the Structural Engineers Association of California, ATC works to transfer state-of-the-art loss-reduction engineering knowledge to the practicing engineering community. Although concentrating to a great extent on earthquake-engineering safety issues, the organization has branched out into engineering challenges associated with other hazards. ATC maintains a longstanding relationship with FEMA, which sponsors its guidance documents and training activities on seismic-performance design guidelines. ATC also engages in knowledge-transfer activities under the sponsorship of a variety of other agencies and entities, including city governments and engineering research consortia. As a boundary organization, ATC plays a distinctive role in using funds provided by such agencies as FEMA to develop guidelines aimed at turning research into practice through direct interactions with the engineering community.

BOX 4.3
The Natural Hazards Center

The Natural Hazards Center (NHC) at the University of Colorado was established in 1976 specifically to address what was then referred to as the knowledge–practice gap in adjustments to natural hazards. Like the other organizations discussed here, the NHC serves as a boundary organization for several segments of the disaster loss-reduction community. Supported by the National Science Foundation and a small group of agencies whose missions center on reducing disaster losses, the NHC engages in a variety of outreach and educational activities that include a newsletter, *The Natural Hazards Observer*; a Web site, hosted chats, and a blog; support for quick-response research; an annual workshop specifically designed to generate interaction among different constituencies in universities, government, international loss-reduction agencies and organizations, and the private sector; a library and information service; dissertation fellowships; and various monographs and special publications. The NHC takes a multidisciplinary approach to disaster resilience, but its strongest constituencies are social-science researchers and the emergency-management community.

such as HSPD-5[13] and HSPD-8,[14] placed overwhelming emphasis on terrorism preparedness, which led to further balkanization of the emergency-management community. In response to federal leadership, states began to develop stand-alone homeland security departments, separate from traditional emergency-management agencies. Concern with terrorism and the accompanying new funding opportunities led to the development of specialized homeland security partnership networks at federal, state, and local levels that were largely independent of networks already established by the traditional emergency-management agencies. Immediate concerns led to effective partnerships that addressed counterterrorism (e.g., joint terrorism task forces), infrastructure protection (e.g., information-sharing and analysis centers—ISACs[15] and Sector Coordinating Councils[16]), and port security (e.g., area security committees). Many such partnerships have been productive, but they tend to depend on federal programs and funding that emphasize collaboration at the national level and do not translate easily to practical local collaboration. Moreover, as noted earlier, networks established for purposes of securing the homeland exhibit chronic problems with information sharing among organizations and levels of government. In addition to generating suspicion on the part of potential partners, information-sharing problems stand in the

[13]See www.fas.org/irp/offdocs/nspd/hspd-5.html (accessed June 30, 2010).

[14]See www.fas.org/irp/offdocs/nspd/hspd-8.html (accessed June 30, 2010).

[15]Eight infrastructure industries were defined as critical to the national economy and well-being by Presidential Decision Directive 63 (PDD 63) in 1998. PDD 63 also proposed the creation of the ISACs (see www.fas.org/irp/offdocs/pdd/pdd-63.htm; accessed June 23, 2010).

[16]See www.dhs.gov/files/partnerships/editorial_0206.shtm (accessed August 6, 2010).

way of the kinds of comprehensive engagement that are needed to develop and nurture resilient collaborative networks.

The fragmentation, inconsistency, and lack of coordination that exist among agencies and entities that have programs and policies in place intended to enhance resilience actually inhibit collaborative efforts. Within the federal family alone, different agencies and subagencies seek to build effective private–public collaboration aimed at coping with hazards, but the efforts are largely uncoordinated. For example, the National Oceanic and Atmospheric Administration and the Environmental Protection Agency sponsor programs aimed at developing local and regional private–public partnerships and supporting decision making in response to climate change and variation and the extreme events that these changes generate (NRC, 2009). The highly successful U.S. Department of Agriculture Cooperative Extension Service[17] and the National Oceanic and Atmospheric Administration National Sea Grant Network[18] are examples of current federal engagement in partnerships to facilitate action and capacity at the community level. FEMA's responsibilities include the development of private–public partnerships for disaster resilience, as do those of its parent agency. The Department of Health and Human Services National Healthcare Facilities Partnership provides funding to improve the surge capacity and disaster preparedness of hospitals and their communities in specific geographic areas through, in part, the strengthening of relationships among the private and public sectors prior to emergencies.[19] The Centers for Disease Control and Prevention are involved in preparedness for all hazards through their Clinician Outreach and Communication Activity (COCA) program that provides up-to-date information to clinicians and two-way communication regarding emerging threats to health.[20]

Many other agencies and offices in DHS, including those charged with infrastructure protection, also aim to achieve resilience goals. One of the 13 divisions of the Obama administration's National Security Council[21] is charged with enhancing the nation's resilience to all threats. The recently released National Security Strategy identifies resilience as one of the nation's top security priorities (The White House, 2010). Similarly, groups such as the International City/County Management Association,[22] the National Governors Association,[23] and the National Association of County and City Health Officials[24] are beginning to focus efforts on issues of resilience, but are not working collaboratively on resilience issues, creating challenges at the community level. Despite this new emphasis and impetus there

[17]See www.csrees.usda.gov/Extension/ (accessed July 1, 2010).
[18]See www.seagrant.noaa.gov/ (accessed July 1, 2010).
[19]See www.phe.gov/preparedness/planning/nhfp/Pages/default.aspx (accessed September 20, 2010).
[20]See emergency.cdc.gov/coca/about.asp (accessed September 20, 2010).
[21]The Obama administration merged the Homeland Security Council and the National Security Council.
[22]See icma.org/en/icma/about/organization_overview (accessed August 27, 2010).
[23]See www.nga.org/portal/site/nga/menuitem.b14a675ba7f89cf9e8ebb856a11010a0 (accessed August 30, 2010).
[24]See /www.naccho.org/ (accessed August 31, 2010).

appears to be little collaboration occurring among all of these organizations, and there is confusion at the community level when private–public collaborative efforts seek information, funding, and other resources.

At the community level, those engaged in collaborative efforts have to be prepared for the lack of coordination among the programs and funding streams intended to support resilience-focused programs; this lack of coordination can lead to conflict and competition among collaborators. Federal funding, often the source of local resilience-focused initiatives, is channeled in narrowly defined programmatic stovepipes. Local area governments try to achieve integrated community goals by using uncoordinated funding streams, such as UASI funds, Public Health Emergency Preparedness funds from the Centers for Disease Control and Prevention,[25] housing funds from the Department of Housing and Urban Development,[26] Coastal Resilience Network funds from the National Oceanic and Atmospheric Administration,[27] and FEMA postdisaster mitigation funds.[28] In such a fragmented and uncoordinated climate, it is understandable that community-level resilience is difficult to generate. That does not mean that resilience-focused private–public collaboration at the community level is not possible and should not occur; rather, communities should be aware of the political climate and strategize accordingly. Private–public collaboration can be the ideal means to leverage disparate federal resources for the benefit of the entire community and thus avoid community-level competition between different sectors.

DEVELOPING METRICS

Few metrics exist to quantify the benefits of collaboration. There is, therefore, little empirical evidence to support funding and policy decisions intended to improve community resilience. An independent study by the MMC attempted to quantify future savings from hazard mitigation activities funded through three major natural hazard mitigation grant programs, including Project Impact, and the results indicated that each dollar spent on FEMA mitigation grants saved society an average of four dollars (MMC, 2005). However, the goals of community resilience building are defined generally as concepts, not as observable and measurable outcomes. The inability to measure and evaluate the outcomes of collaboration makes it more difficult for organizations and individuals to commit to collaborative solutions.

Case studies like those associated with Project Impact tend to be anecdotal. Longitudinal data are seldom collected, and confounding variables that are linked to outcomes are not

---

[25]See www.bt.cdc.gov/planning/ (accessed June 30, 2010).

[26]See portal.hud.gov/portal/page/portal/HUD/program_offices/administration/grants/fundsavail (accessed June 30, 2010).

[27]See www.csc.noaa.gov/funding/ (accessed August 9, 2010).

[28]See www.fema.gov/government/grant/hma/index.shtm (accessed June 30, 2010).

identified (Magsino, 2009; NRC, 2010). It is difficult to get many people and organizations to commit to a process when the destination is not known and effective means of measuring progress do not exist. People respond best when outcomes can be observed and measured. Businesses are reluctant to commit when the costs of collaboration are clear but the benefits are not. Participating in collaboration is often a function of individual commitment and willingness to accept risk. In most organizations, however, people are rewarded only for activities that are measured, so individual success in building essential collaboration is typically unrewarded.

Communities will probably not have the resources to develop the kinds of metrics needed for quantitative evaluation of increases in resilience and similar factors resulting from their resilience-focused private–public collaborations. Until research shows how such outcomes can be measured, communities can develop goals and mechanisms to meet them that include discreet milestones to describe the effectiveness of collaboration. Such descriptions may not completely quantify the outcomes for funding or policy-development purposes, but they can keep high or raise enthusiasm for engagement.

REFERENCES

ASIS International. 2009. SPC.1-2009 Organizational Resilience Standard Adopted by the DHS in PS-Prep. Alexandria, VA. Available at www.asisonline.org/guidelines/or.xml (accessed June 30, 2010).

BSI Group. 2009. BS 25999 Business continuity. London, UK: British Standards Institution. Available at www.bsigroup.com/en/Assessment-and-certification-services/management-systems/Standards-and-Schemes/BS-25999/ (accessed June 30, 2010).

Colten, C. E., R. W. Kates, and S. B. Laska. 2008. Community Resilience: Lessons from New Orleans and Hurricane Katrina. Oak Ridge, TN: Community & Regional Resilience Institute. Available at www.rwkates.org/pdfs/a2008.03.pdf (accessed August 31, 2010).

Cutter, S. L., L. Barnes, M. Berry, C. Burton, E. Evans, E. Tate, and J. Webb. 2008. A Place Based Model for Understanding Community Resilience to Natural Disasters. *Global Environmental Change* 18(4): 598-606.

Elder, K., S. Xirasagar, N. Miller, S. A. Bowen, S. Glover, and C. Piper. 2007. African Americans' Decisions not to Evacuate New Orleans Before Hurricane Katrina: A Qualitative Study. *American Journal of Public Health* 97(S1): 124-129.

Flynn, S. E., and D. B. Prieto. 2006. Mobilizing the Private Sector to Support Homeland Security. Washington, DC: Council on Foreign Relations. Available at www.cfr.org/publication/10457/neglected_defense.html (accessed June 30, 2010).

GAO (Government Accountability Office). 2005. Clear Policies and Oversight Needed for Designation of Sensitive Security Information. GAO-05-677. Washington, DC. Available at www.gao.gov/new.items/d05677.pdf (accessed August 4, 2010).

GAO (Government Accountability Office). 2008. Definition of the Results to be Achieved in Terrorism-Related Information Sharing is Needed to Guide Implementation and Assess Progress. GAO-08-637T. Washington, DC. Available at www.gao.gov/new.items/d08637t.pdf (accessed August 4, 2010).

Guston, D. H. 1999. Stabilizing the boundary between U.S. politics and science: The role of the Office of Technology Transfer as a boundary organization. *Social Studies of Science* 29(1): 87-112.

Guston, D. H. 2000. *Between Politics and Science: Assuring the Integrity and Productivity of Research.* New York: Cambridge University Press.

Guston, D. H. 2001. 'Boundary organizations' in environmental policy and science: An introduction. *Science, Technology, and Human Values* 26: 399-408.

Magsino, S. 2009. *Applications of Social Network Analysis for Building Community Disaster Resilience: Workshop Summary.* Washington, DC: The National Academies Press.

MMC (Multihazard Mitigation Council). 2005. Natural Hazard Mitigation Saves: An Independent Study to Assess the Future Savings from Mitigation Activities. Washington, DC: National Institute of Building Sciences. Available at www.nibs.org/index.php/mmc/projects/nhms/ (accessed June 30, 2010).

NFPA (National Fire Protection Association). 2007. NFPA 1600: Standard on Disaster/Emergency Management and Business Continuity Programs. Available at www.nfpa.org/assets/files/pdf/nfpa1600.pdf (accessed June 30, 2010).

NRC (National Research Council). 2006. *Facing Hazards and Disasters: Understanding Human Dimensions.* Washington, DC: The National Academies Press.

NRC (National Research Council). 2009. *Observing Weather and Climate from the Ground Up: A Nationwide Network of Networks.* Washington, DC: The National Academies Press.

NRC (National Research Council). 2010. *Private-Public Sector Collaboration to Enhance Community Disaster Resilience: A Workshop Report.* Washington, DC: The National Academies Press.

Public Law 110-53. 2007. Title IX – Private Sector Preparedness. Implementing the 9/11 Commission Recommendations Act of 2007. August 3. Available at www.nemronline.org/TITLE IX Private Sector Preparedness.pdf (accessed June 30, 2010).

Tierney, K. J. 2007. From the margins to the mainstream? Disaster research at the crossroads. *Annual Review of Sociology* 33: 503-525.

Townsend, F. F. 2006. Federal Response to Hurricane Katrina: Lessons Learned. Washington, DC: Washington Government Printing Office. February. Available at georgewbush-whitehouse.archives.gov/reports/katrina-lessons-learned/ (accessed June 30, 2010).

The White House. 2010. *National Security Strategy*, available at www.whitehouse.gov/sites/default/files/rss_viewer/national_security_strategy.pdf (accessed August 6, 2010).

CHAPTER FIVE

Research Opportunities

Successful collaborative efforts to create disaster-resilient communities will take into account motivators of and inhibitors to forming partnerships, sustaining them, and gaining knowledge about partner roles. Because such collaborative work is in its nascent stages in much of the nation and because social change and vulnerability to hazards are evolving so rapidly, parallel programs of collaboration and research are imperative.

Key topics for research include:

- How, when, and why collaboration works or fails.
- Ways of accounting for different outcomes that result from alternative partnership-building strategies, such as bottom-up voluntary collaboration vs. partnerships and partnership-building strategies initiated or funded by government.
- Predicting partnership legitimacy, effectiveness, mainstreaming, and institutionalization.
- Appropriate metrics for quantifying the costs and benefits resulting from investments in collaboration and resilience-building efforts.

Those issues need to be understood and evaluated for a variety of communities and threats, and results of the issues documented, archived, and disseminated.

Research findings can inform training strategies as well as new program areas, and can lead to the development of more refined conceptual frameworks. Furthermore, private–public sector collaboration could be improved with a better understanding of how such collaboration is born, develops, and functions in the larger context of state-level and federal-level initiatives and, where applicable, in the larger context of global businesses and national and international civil society. The committee discusses below a set of research initiatives that could be targeted for investment by the Department of Homeland Security (DHS) and others interested in deepening knowledge on the topic of resilience-focused private–public sector collaboration.

BUSINESS-SECTOR MOTIVATORS

Investigate factors most likely to motivate businesses of all sizes to collaborate with the public sector to build disaster resilience in different types of communities (for example, rural and urban).

As described in Chapter 4 and in the summary of the committee's workshop (NRC, 2010), there are a number of impediments to business participation in private–public collaboration of all types, including those centering on disaster resilience. The barriers include private–public sector cultural differences, concerns about information sharing, and wariness of government mandates and regulations. What is not clear is how to overcome such challenges and increase incentives for business participation in disaster-loss–reduction activities. Incentives are multifaceted and vary among different types of businesses. For example, some workshop participants argued that business-sector involvement in private–public sector collaboration is motivated partially by an understanding of the direct benefits of participation in resilience-building collaboration, the desire to maintain favorable public perceptions, and liability concerns. The bulk of what is known about factors that motivate business engagement in resilience-enhancing activities is anecdotal. It is impossible to answer even simple questions, such as whether business organizations are motivated primarily by concerns about the safety of their own properties and operations; or whether business size, profitability, length of tenure in a community, being a branch or franchise of a larger national organization, or participating in other community-improvement ventures predicts business involvement in disaster-related private–public partnerships.

Anecdotal evidence points to the need to understand better the diverse views of resilience and emergency-management issues held by both private-sector and public-sector collaborators. As described in Chapter 3, public-safety agencies often underestimate private-sector interest and involvement in emergency-preparedness efforts. Similarly, private-sector groups often overestimate the capabilities of public-sector partners, failing to recognize the need for their own contributions to disaster management. Research to assess the effect of community context on private-sector participation is also needed: Is such participation more likely in rural communities than in urban or suburban communities? In higher-risk communities as opposed to those in which disasters are less frequent?

In sum, there is a pressing need to understand better why the business sector is drawn to disaster-resilience–building collaboration at the community level, the different perceptions held by various collaborators, and the types of incentives that resonate with business leaders. Furthermore, partnerships and partner motivations are not static over the long term. There are different levels of collaboration, ranging from simple networking to forming contractual partnerships. What incentives are most likely to encourage greater levels of participation?

Research on such issues can help in designing viable partnership models and guidelines appropriate for a variety of business types and sizes.

INTEGRATING NONGOVERNMENTAL ORGANIZATIONS

Focus research on how to motivate and integrate community-based, faith-based, and other nongovernment organizations—including those not crisis oriented—into resilience-focused collaboration.

Community-based organizations, faith-based organizations (FBOs), and other nongovernment organizations (NGOs) play critical roles in all phases of the disaster cycle. They help develop social capital, serve and represent disenfranchised community members, and in general provide a social safety net for diverse at-risk populations. It is to such organizations that vulnerable community residents will turn when disaster strikes. For that reason alone, communities have a vested interest in the involvement of the nonprofit sector in resilience-enhancing activities. However, the small amount of existing research on how resilient this sector is indicates that NGOs are not well prepared for disasters and that representatives of community-based organizations are rarely involved in community disaster-resilience efforts (Drabek, 2003). One of the biggest gaps in our knowledge is the role played by what is referred to as understudied "noncrisis-relevant nonprofits" and community-based organizations (for example, homeless shelters, agencies serving immigrants, and community clinics). Such targeted research can help communities identify their unknown or neglected facilitators during times of disaster. Understanding how to act on that knowledge could provide a means of empowering those groups to operate most efficiently for their own benefit and for the benefit of the community as a whole. More cost-effective and viable mobilization strategies, particularly for communities in perpetual states similar to disaster because of such conditions as extreme poverty, could be identified. Furthermore, research on how partnership agendas can be reframed to be more inclusive may help bring in important but overlooked community stakeholders.

CHANGING EMERGENCY-MANAGEMENT CULTURE

Focus research on how the emergency-management and homeland security sectors can be moved toward a "culture of collaboration" that engages the full fabric of the community in enhancing resilience.

Findings discussed in this report indicate that private-sector organizations including NGOs have difficulty forming private–public partnerships, and government agencies charged with emergency-management responsibilities face similar barriers. The question of how best to move government emergency-management and homeland security agencies toward a culture of collaboration has received little research attention (for more discussion, see Stanley and Waugh, 2001; Drabek, 2003; McEntire, 2007). Research is therefore needed to explore ways to overcome structural, cultural, educational, training, and other barriers that may prevent those in the public emergency-management sector from adopting more collaborative models for resilience enhancement.

As discussed in Chapter 4, those in the government emergency-management and homeland security agencies tend to be unfamiliar with the concerns and perspectives that typify the private, including the nonprofit, sector. They may be unfamiliar with the kinds of activities and processes needed to initiate and nurture cross-sector collaboration. Many entities and personnel in emergency management have yet to embrace the concept of collaborative emergency management even though the concept is nearly two decades old. Some government agencies and personnel remain more comfortable with top-down "command and control" frameworks than with approaches that emphasize collaboration and network management. Such perspectives are probably rooted in earlier training and professional experiences—for example, in the military or law enforcement—or in concerns about homeland security. They may also be rooted in lack of knowledge about the role of civil society in disaster management, in concerns about "turf" and organizational prerogatives, and possibly even in generational differences. Researchers find the National Response Framework (FEMA, 2008) more "collaboration-friendly" than earlier plans for intergovernment disaster response (see, for example, discussions in Gazley et al., 2009), but the fact remains that a "culture of collaboration" has not yet taken hold in the emergency-management and homeland security communities.

BUILDING SOCIAL CAPITAL

Focus research on ways to build capacity for resilience-focused private–public sector collaboration.

Research on disaster resilience has focused increasingly on the relationship between social capital and resilience with an emphasis on social capital as the foundation for community adaptive capacity (see Norris et al., 2008). The formation of effective and productive social networks constitutes a key element in the development of social capital, and private–public partnerships can provide an infrastructure for such networks. Recognizing the importance of social capital and capacity building raises the need for research on capacity-building strategies, including studies that focus on the kinds of training needed for leaders in the private and public sectors; on how collaboration skill sets are built at the community level; and on how creativity and innovation can be fostered within collaboration, for example, by tapping into the potential that is inherent in new information and communication technologies. Some workshop participants spoke of the need for peer mentoring as a capacity-building strategy, but other strategies, such as personnel exchanges across sectors and new training experiences for government officials, also need to be assessed.

LEARNING THROUGH SUPPORT OF COLLABORATION

Focus on research and demonstration projects that quantify risk and outcome metrics, enhance disaster resilience at the community level, and document best practices.

New efforts to support and nurture community-level resilience-focused private–public collaboration could include research and demonstration projects aimed at enhancing disaster resilience at the community level and documenting best practices. Project Impact was a major federal government initiative whose goal was the development of local partnership networks for risk and vulnerability assessments, disaster-mitigation projects, and public education (Witt and Morgan, 2002). In Project Impact, the federal government set general guidelines and provided funding to local communities, but it did not mandate how local programs should be organized, nor did it attempt to micromanage local project activities. The Federal Emergency Management Agency, which funded Project Impact, also funded a series of formative evaluation studies whose findings are discussed in Chapter 3 of this report. The studies documented many aspects of program operations, including how programs were organized, the activities that were undertaken under the rubric of Project Impact, and the kinds of partnerships that were developed.

Recognizing that private–public partnership and broad community mobilization are needed to improve the disaster resilience of communities, DHS might sponsor a series of research and demonstration projects across the nation. The new projects could fully integrate research and practice, beginning with the initial phase of project development, and could be conceptualized as living laboratories that provide opportunities for both researchers and practitioners. Research could be designed and undertaken with the explicit goal of documenting the effectiveness of collaboration, the costs and benefits to collaborators, and the metrics for these variables. Both process- and outcome-related variables could be addressed. Longitudinal and comparative designs could be key elements in the research and demonstration projects.

Continuing collaboration between researchers and practitioners in diverse local resilience-building efforts offers the potential for the application of principles of adaptive management—applied widely in programs that address environmental problems other than disasters (Walters, 1986; Lee, 1993; Wise, 2006). Using an adaptive-management approach, program participants and their research collaborators determine measures to undertake, implement the measures, assess their effects, learn how to improve on the basis of the assessments, adjust programs accordingly, and then continue with the cycle of implementation, assessment, learning, and program adjustment. The systematic use of an adaptive-management approach can improve programs continually and shed light on best practices and strategies for achieving resilience objectives.

Focus on research and related activities that produce comparable nationwide data on both vulnerability and resilience.

Various approaches are being used to assess the nation's vulnerability to disasters and other hazards. HAZUS and HAZUS-MH,[1] for example, are widely used vulnerability-assessment tools. Since its inception, DHS has been engaged in diverse activities involving multiple programs and directorates to quantify risks to the nation's critical infrastructure and to compare risk and vulnerability in U.S. communities. Researchers have developed various measures of social vulnerability; the most widely recognized is the Social Vulnerability Index (SOVI),[2] developed by Susan Cutter and other researchers at the University of South Carolina. Activities are also under way to assess community resilience with various measures.

Despite the progress made in those and related fields, the nation lacks an agreed-on set of vulnerability and resilience indicators that would make it possible to measure and assess them in communities and over time. Without such measures, it will be impossible to gauge progress in efforts to improve resilience or to compare community progress. That need has been recognized in the past and, with funding from the National Science Foundation (NSF) and the U.S. Geological Survey (USGS), hazard and disaster researchers met at a workshop in June 2008 and developed a plan and set of research recommendations for the development of a Resiliency and Vulnerability Observatory Network, or RAVON (Peacock et al., 2008). The goal of RAVON would be to systematize the collection, retention, and dissemination of data that are relevant to the measurement of vulnerability and resilience. It would incorporate other key indicators, such as those related to risk assessment, perception, and management; hazard mitigation; and disaster recovery and reconstruction. As envisioned by the 2008 workshop participants, RAVON would combine the best elements of virtual and place-based activities and research–practitioner collaboration and would borrow elements of similar existing activities, such as the Long Term Ecological Research Network[3] (LTER) and the National Ecological Observatory Network[4] (NEON).[5]

To understand the extent to which the nation is moving toward a more resilient and less vulnerable future, and to understand the factors affecting that movement, reliable, valid, and

[1]See www.fema.gov/plan/prevent/hazus/ (accessed July 1, 2010).

[2]See webra.cas.sc.edu/hvri/products/sovi.aspx (accessed July 1, 2010).

[3]See www.lternet.edu/ (accessed July 1, 2010).

[4]See www.neoninc.org/ (accessed July 1, 2010).

[5]For more detailed discussions of the proposed RAVON activities and organizational structure, see Peacock et al. (2008).

systematically collected indicators are essential. Sponsoring a network, such as RAVON, is consistent with the mission of DHS; indeed, it is difficult to envision how such a network could be developed without substantial input on the part of DHS.

A REPOSITORY OF INFORMATION

Communities around the nation have few information resources on collaboration for developing community disaster resilience. Information and guidance exist but are scattered throughout the peer-reviewed literature, government reports, research-project reports, and organizational and institutional Web sites; this makes access difficult for the average local agency or business. As sponsored research on private–public collaboration develops and matures, DHS itself will need a means of disseminating research findings.

Establish a national repository and clearinghouse, administered by a neutral entity, to archive and disseminate information on community resilience-focused private–public sector collaboration models, operational frameworks, community disaster-resilience case studies, evidence-based best practices, and resilience-related data and research findings. Relevant stakeholders in all sectors and at all levels should convene to determine how to structure and fund this entity.

Workshop (NRC, 2010) and committee discussions have revealed that nongovernment partners are likely to prefer information and guidance from third-party sources that are considered independent and disinterested. That finding and a recognition of the importance of "boundary organizations" (Guston, 2001) in bridging the research-policy–practice gap, form the basis of this recommendation.

Tentatively called the Center for Best Practices in Disaster Resilience, the entity or network of entities would provide information and guidance free of charge and in formats that are readily accessible and comprehensible by private-sector and public-sector leaders, emergency-management practitioners, and researchers. It would make available a variety of products, from peer-reviewed publications to existing and emerging "toolkits" for those engaged in private–public collaboration. As an NGO, it would serve as the "honest broker" and facilitator for private–public sector interactions on resilience issues.

In considering the need for an independent repository of information and expertise, the committee stopped short of offering advice on how such a resource should be structured and funded. Those are decisions best made by a broad-based and trusted coalition of public- and private-sector stakeholders and experts in the delivery of guidance information to diverse users. Agencies that support research and practice in community disaster resilience (NSF, DHS, the National Oceanic and Atmospheric Administration, USGS, and others) have an important role in making these decisions, but the committee concludes that the resource should not be perceived as "owned" by any one agency. Broad-based participation is critical to ensure the legitimacy and long-term viability of the center, just as the committee has shown it is critical in community-based, resilience-focused private–public collaboration.

FINAL THOUGHTS

The term *resilience* was not in use when the National Governors Association developed its comprehensive emergency-management guide in 1979 (NGA, 1979). That document was written to assist governors in the transition to all-hazards approaches to emergency management through all phases of the disaster cycle. It emphasized coordination of resources and knowledge and the state's supporting role in disaster response after primary response by local governments. Many of the conclusions reached by the present National Research Council committee are similar to those reached over 30 years ago in the report to the governors but scaled down to the community level, broadened to include a much more active role of the private sector, and made applicable with advances in communication technology. Our ability to identify, analyze, tap into, and create communication networks far exceeds what the governors in 1979 may have imagined. Our ability to listen, engender trust, and collaborate, however, has not kept up with our ability to transmit messages. To create a resilient nation, a nation of resilient communities must be created. Resilient communities can be and are being created through resilience-focused private–public collaboration originating in the community at the grassroots level and including representatives of all segments of the community with facilitation and coordinated support from higher levels of government and the private sector.

In reading a report like the present one—beginning with Secretary of Homeland Security Napolitano's remarks and continuing to the last guideline—it is natural to see building disaster-resilient communities as an end unto itself. But the stark reality is that the United States is attempting to maintain and foster its entire national agenda—to provide for public safety and health, to grow the economy, to protect the environment, and to maintain basic human values of freedom and dignity—community by community. And our nation of communities seeks to accomplish those goals on a planet that moves its physical matter from one place to another through extreme events (e.g., earthquakes, volcanic eruptions, hurricanes). Building disaster-resilient communities is essential for the whole of our national hopes and aspirations. Private–public collaboration is the starting point for building such resilience.

REFERENCES

Drabek, T. E. 2003. *Strategies for Coordinating Emergency Responses*. Boulder, CO: University of Colorado, Institute of Behavioral Science.

FEMA (Federal Emergency Management Agency). 2008. National Response Framework. Washington, DC: U.S. Department of Homeland Security. Available at www.fema.gov/pdf/emergency/nrf/nrf-core.pdf (accessed March 11, 2010).

Gazley, B., J. L. Brudney, and D. Schneck. 2009. "Using risk indicators to predict collaborative emergency management and county emergency preparedness." Paper prepared for presentation at the meeting of the Public Management Research Association, Columbus, OH, Oct. 1. [Note: permission has been granted to cite this reference.]

Guston, D. H. 2001. "'Boundary organizations' in environmental policy and science: An introduction." *Science, Technology, and Human Values* 26: 399-408).

Lee, K. N. 1993. Compass and Gyroscope: Integrating Science and Politics for the Environment. Washington, DC: Island Press.

McEntire, D. 2007. *Disaster Response and Recovery*. Hoboken, NJ: John Wiley and Sons.

NGA (National Governors Association). 1979. Comprehensive Emergency Management: A Governor's Guide. Washington, DC. Available at training.fema.gov/EMIWeb/edu/docs/Comprehensive%20EM%20-%20NGA.doc (accessed June 20, 2010).

Norris, F. H., S. P. Stevens, B. Pfefferbaum, K. F. Wyche, and R. L. Pfefferbaum. 2008. Community resilience as a metaphor, theory, set of capacities, and strategy for disaster readiness. *American Journal of Community Psychology* 41(1-2):127-150.

NRC (National Research Council). 2010. *Private-Public Sector Collaboration to Enhance Community Disaster Resilience: A Workshop Report.* Washington, DC: The National Academies Press.

Peacock, W. G., H. Kunreuther, W. H. Hooke, S. L. Cutter, S. E. Chang, and P. R. Berke. 2008. Toward a Resiliency and Vulnerability Observatory Network: RAVON. College Station, TX: Hazard Reduction and Recovery Center, Texas A&M University. HRRC report 08-02-R.

Stanley, E. and W. L. Waugh Jr. 2001. Emergency managers for the new millennium. In *Handbook of Crisis and Emergency Management*, A. Farzimand (ed.), New York: Marcel Dekker.

Walters, C. J. 1986. Adaptive Management of Environmental Resources. Caldwell, NJ: Blackburn Press.

Wise, C. R. 2006. "Organizing for homeland security after Katrina: Is adaptive management what's missing?" *Public Administration Review* 66: 302-318.

Witt, J. L., and J. Morgan. 2002. *Stronger in the Broken Places: Nine Lessons for Turning Crisis into Triumph*. New York: Henry Holt & Company.

Appendixes

Committee Biographies

William H. Hooke is a Senior Policy Fellow and the Director of the Atmospheric Policy Program at the American Meteorological Society (AMS) in Washington, DC. Prior to arriving at AMS in 2000, Dr. Hooke worked for the National Oceanic and Atmospheric Administration (NOAA) and antecedent agencies for 33 years. After six years of research with NOAA he moved into a series of management positions of increasing scope and responsibility including Chief of the Wave Propagation Laboratory Atmospheric Studies Branch, Director of NOAA's Environmental Sciences Group (now the Forecast Systems Lab), Deputy Chief Scientist, and Acting Chief Scientist of NOAA. Between 1993 and 2000, he held two national responsibilities: Director of the U.S. Weather Research Program Office, and Chair of the Interagency Subcommittee for Natural Disaster Reduction of the National Science and Technology Council Committee on Environment and Natural Resources. Dr. Hooke was a faculty member at the University of Colorado from 1967 to 1987, and served as a fellow of two NOAA Joint Institutes (CIRES, 1971-1977; CIRA 1987-2000). The author of over fifty refereed publications and co-author of one book, Dr. Hooke holds a B.S. (Physics Honors) from Swarthmore College (1964), and S.M. (1966) and Ph.D. (1967) from the University of Chicago. He recently chaired the NAS/NRC Disasters Roundtable, and was elected a member of the American Philosophical Society in 2006.

Arrietta Chakos is public policy advisor on urban disaster resilience. She most recently served as project director of the Harvard Kennedy School's Acting in Time Disaster Recovery Research Project and previously served as assistant city manager in Berkeley, California, directing intergovernmental coordination and innovative hazard mitigation initiatives. She works on disaster risk reduction public policy and sustainable community engagement. The Acting in Time Disaster Recovery Research Project focused on identifying effective social and government interventions to reduce disaster risk and by supporting communities to responsibly implement safety measures. Ms. Chakos has worked with the Federal Emergency Management Agency (FEMA) and the California's Office of Emergency Services on hazard mitigation programs. She has served as a technical advisor for FEMA on risk mitigation and disaster loss estimation. She has advised GeoHazards International; the Organization for Economic Cooperation and Development (OECD); the World Bank; California's Office of Emergency Services; the Association of Bay Area Governments; and the Center for BioSecurity on disaster and community engagement issues. She has been an invited speaker at the Disasters' Roundtable at the National Academies; the 2006 conference on the 1906 San Francisco earthquake; and, the Natural Hazards' Center annual conference at the University of Colorado. International efforts include U.S./Japan and U.S./China seismic workshops, at the UN hazards conference on the Kobe earthquake and recent academic and technical conferences in China examining seismic safety. Publications include papers on disaster issues for numerous technical conferences on disaster risk reduction; for the American Society of Civil Engineers; for Spectra, a professional publication; and for the Natural Hazards' Observer. She contributed a chapter to OECD's book, Keeping Schools Safe in Earthquake Country and to Global Warming, Natural Hazards, and Emergency Management (2008). Ms. Chakos received her B.A. in English from California State University, Humboldt, and her M.P.A. from Harvard University Kennedy School.

Ann-Margaret Esnard is a professor and director of the Visual Planning Technology Lab (VPT Lab) in Florida Atlantic University's School of Urban and Regional Planning. Dr. Esnard's expertise encompasses coastal vulnerability assessment, GIS/spatial analysis, displacement vulnerability, disaster planning, and land-use planning. She has been involved in a number of related research initiatives, including two NSF-funded projects on topics of hurricane-related population displacement, and long-term recovery. She has written on topics that include: population displacement from catastrophic hurricanes, vulnerability assessments of coastal and flood hazards, quality of life and holistic disaster recovery geospatial technologies, GIS education, public participation GIS, and environmental justice. Esnard has served on a number of local, state and national committee including: the Steering Committee for Evaluation of the National Flood Insurance Program, the Disasters Roundtable of the National Research Council; and the State of Florida Post-Disaster Redevelopment Planning initiative. Dr. Esnard holds degrees in Agricultural Engineering (B.Sc., University of the West Indies-Trinidad), Agronomy and Soils (M.S., University of Puerto Rico-Mayaguez) and Regional Planning (Ph.D., UMASS-Amherst). She also completed a two-year postdoc at UNC-Chapel Hill.

John (Jack) R. Harrald is a research professor and adjunct professor of public policy at the Virginia Polytechnic Institute and State University Center for Technology, Security, and Policy. Dr. Harrald is Co-Director Emeritus of The George Washington University

(GWU) Institute for Crisis, Disaster, and Risk Management; Professor Emeritus of Engineering Management and Systems Engineering in the GWU School of Engineering and Applied Science; and Chairman of the National Research Council Disasters Roundtable Steering Committee. He is cofounder and Executive Editor Emeritus of the electronic *Journal of Homeland Security and Emergency Management.* He is the immediate past president of The International Emergency Management Society and former associate director of the National Ports and Waterways Institute. During his 22-year career as a U.S. Coast Guard officer, he has also worked as a practitioner, retiring in the grade of Captain. He has written and published in the fields of crisis management, emergency management, management science, risk and vulnerability analysis, and maritime safety. Dr. Harrald received his B.S. in Engineering from the U.S. Coast Guard Academy, a M.A.L.S. from Wesleyan University, a M.S. from the Massachusetts Institute of Technology where he was an Alfred P. Sloan Fellow, and an MBA and Ph.D. from Rensselaer Polytechnic Institute.

Lynne Kidder is a senior advisor to the Center for Excellence in Disaster Management and Humanitarian Assistance (COE-DMHA)—a Department of Defense organization focused on improved civil–military interagency coordination, capacity building, and resilience-focused public–private collaboration in the context of disaster response and humanitarian assistance. From 2005–2010, she served as vice president and senior vice president for Partnerships at Business Executives for National Security (BENS), where she directed BENS's national program to facilitate community and statewide public–private partnerships toward regional all-hazards disaster resilience. While at BENS, Ms. Kidder convened a coalition of national business leaders, professional/trade organizations, academics, NGOs, military, and agency partners, to propose a national mechanism to strengthen public-private collaboration at all levels of government. She is the former executive director of the North Bay Council, a nonprofit organization of C-level executives in Northern California, where she implemented numerous initiatives between private employers and state and local officials. Ms. Kidder's previous professional experience includes eight years as professional staff in the U.S. Senate, executive-level management in state government, and corporate government affairs. She holds a B.A. from Indiana University (College of Arts and Sciences), a Master's degree from the University of Texas at Austin, and did additional post-graduate study in public administration at George Mason University. Ms. Kidder is the co-chair of the Institute of Medicine's Forum on Medical and Public Health Preparedness for Catastrophic Events and serves on numerous boards and advisory committees pertaining to resilience-focused public-private collaboration and all-hazards preparedness.

Michael T. Lesnick is cofounder of and senior partner at Meridian Institute, a nonprofit organization that provides neutral conflict management and multistakeholder collaborative problem solving services domestically and internationally. Dr. Lesnick has over 30 years of experience designing and managing multiparty information sharing, problem solving and conflict management processes. His work with decision makers and stakeholders from government, corporations, nongovernmental organizations, international institutions, and scientific bodies has resulted in bringing practical solutions and new public–private partnerships to some of society's most controversial and complex problems particularly in the areas of national and homeland security, environment and sustainable development, public health, food security, climate change, international development, and science policy. Dr. Lesnick facilitated the White House Hurricane Katrina Stakeholder Summit as well as interagency and stakeholder processes in the development of the National Infrastructure Protection Plan and the National Response Framework. He directed projects that resulted in the formation of nine critical infrastructure and key resource sector coordinating councils at the national level as well as pandemic planning processes for the DHS Office of Infrastructure Protection. Dr. Lesnick works extensively with the Community and Regional Resilience Institute (CARRI). He has been the project director of over 100 domestic and international multistakeholder collaboration processes. He has published in the areas of facilitation, mediation and strategy assessment. He holds an M.S. and Ph.D. from The University of Michigan where he was also a postdoctoral fellow in Environment and Collaborative Problem-Solving and Conflict Management.

Inés Pearce is Chief Executive of Pearce Global Partners (PGP), addressing the needs of government, business, nonprofits and communities to reduce the potential for loss of life and property from natural and human-caused disasters. Ms. Pearce is a business continuity and emergency management expert with 17 years of professional experience, including 12 years with public–private partnerships. She also serves as the Senior Disaster Response Advisor for the Business Civic Leadership Center (BCLC) of the U.S. Chamber of Commerce, where she is BCLC's primary point of contact for community-level disaster preparedness, recovery, and partnership coordination. She has also served as a U.S. Chamber of Commerce liaison during disasters to facilitate long-term recovery, such as 2008s flooding in Iowa, storms in Florida, and hurricanes in Texas and Louisiana. Before launching PGP, Ms. Pearce was appointed as Seattle Project Impact Director for the City of Seattle Emergency Management, managing four mitigation programs that provided resources for safer schools, homes, and businesses, as well as better hazard maps. During her tenure, Seattle Project Impact received numerous national excellence awards. As an expert in public–private partnerships, Ms. Pearce has represented the World Economic Forum at the United Nations' (UN) Global Platform for Disaster Risk Reduction in Geneva, Switzerland, and has addressed the UN regarding public–private partnerships at the World Conference for Disaster Reduction in Kobe, Japan. In 2003, Ms. Pearce was inducted into the Contingency Planning & Management (CPM) Hall of Fame in the Public Servant Category. She has also received two National Excellence Awards from the Western States Seismic Policy Council, and in

2009, received an Award of Recognition from the City of Los Angeles for the successful planning of the Great Southern California ShakeOut, the largest earthquake drill in U.S. history with 5.5 million participants. Ms. Pearce is President of the Contingency Planning & Recovery Management (CPARM) group, the Disaster Resistant Business (DRB) Toolkit Workgroup, and on the Board of CREW, the Cascadia Regional Earthquake Workgroup. She received her B.A. degree in political science from Gonzaga University.

Randolph H. Rowel is an assistant professor and Director of the Why Culture Matters Disaster Studies Project at the Morgan State University School of Community Health and Policy. Dr. Rowel has over 25 years experience in community health education with considerable expertise in community organizing and empowerment, partnership development, and social marketing. He teaches Community Needs and Solutions, Community-Based Participatory Research, and Qualitative Research in Public Health and has been an invited presenter at numerous emergency management related conferences to speak on community engagement and the cultural implications of disasters. Dr. Rowel serves as an investigator for the Department of Homeland Security funded National Center for Preparedness and Catastrophic Event Response (PACER), where he is conducting studies to examine the relationship between daily crisis and preparedness behavior and community engagement strategies for low-income populations. As a PACER investigator, Dr. Rowel is also developing culturally appropriate disaster preparedness curriculum for faith-based leaders. In partnership with Maryland Department of Health and Mental Hygiene, Dr. Rowel recently completed a project that examined knowledge, perceptions, and natural disaster experiences of low-income African American and Spanish-speaking Latino populations. This initiative led to publishing a "Guide to Enhance Grassroots Risk Communication Among Low-Income Populations" which provides practical, step-by-step instructions on how to work with grassroots organizations in order to deliver critical information to low-income populations before, during, and after a disaster. Dr. Rowel recently served on National Academies Ad Hoc Committee to plan a Social Network Analysis (SNA) workshop. The workshop examined the current state of the art in SNA and its applicability to the identification, construction, and strengthening of networks within U.S. communities for the purpose of building community resilience. He received his undergraduate degree at Morgan State University and his masters and doctoral degrees from the University of Utah and the University of Maryland College Park, respectively.

Kathleen J. Tierney is professor of sociology and director of the Natural Hazards Center at the University of Colorado at Boulder. The Hazards Center is housed in the Institute of Behavioral Science, where she holds a joint appointment. Dr. Tierney's research focuses on the social dimensions of hazards and disasters, including natural, technological, and human-induced extreme events. She is co-author of *Disasters, Collective Behavior and Social*

Organization (1994) and *Facing the Unexpected: Disaster Preparedness and Response in the United States* (2001) and co-editor of *Emergency Management: Principles and Practice for Local Government* (2007). Dr. Tierney has published widely on hazards- and disaster-related topics in such publications as the *Annual Review of Sociology*, the *Annals of the American Academy of Political and Social Science*, *Contemporary Sociology*, *Sociological Spectrum*, *Sociological Forum*, the *Journal of Homeland Security and Emergency Management*, and other journals. She has served as a member of the National Academies Committee on Disaster Research in the Social Sciences, the Panel on Strategies and Methods for Climate-Related Decision Support, and the "America's Climate Choices" Panel on Informing an Effective Response to Climate Change. Over the course of her career, she has held research and faculty positions at the University of California, Los Angeles, the University of Southern California, the University of California at Irvine, and the University of Delaware. Dr. Tierney earned a Ph.D. in Sociology from The Ohio State University.

Brent H. Woodworth is currently President and CEO of Los Angeles Emergency Preparedness Foundation. He is a well-known leader in domestic and international crisis management with a distinguished history of working in partnership with government agencies, private sector companies, academic institutions, faith-based organizations, and nonprofits. In December 2007 he took his retirement from IBM Corporation after 32 years of service which included the development and management of all worldwide crisis response team operations. Over the past several years, Mr. Woodworth has led his response team in response to over 70 major natural and man-made disasters in 49 countries. Mr. Woodworth's domestic response efforts include the 1992 Civil Unrest in Los Angeles followed by the 1994 Northridge Earthquake, Oklahoma City Bombing, 9/11 World Trade Center attacks, Hurricane Katrina, and multiple flooding, wind, fire, and seismic events. In 1998, Mr. Woodworth was appointed by FEMA Director James Lee Witt to serve on a U.S. Congressional designated committee where he co-authored the national plan for predisaster mitigation. Mr. Woodworth has served on national and local committees and boards including the National Institute of Building Sciences (NIBS) board of directors; the U.S. Multihazard Mitigation Council (MMC) as chairman; the Advisory Committee on Earthquake Hazards Reduction (ACEHR) board of directors; and as the Los Angeles Emergency Preparedness Foundation president and CEO. Mr. Woodworth is the recipient of multiple industry awards and a well published author on disaster preparedness, public–private partnerships, and crisis events. One example of Mr. Woodworth's public–private sector collaboration focus includes his successful negotiation with Starbucks Corporation and T-Mobile, Inc., whereby they provided free wireless connection service at over 1000 locations from Santa Barbara to the U.S.–Mexico border during the California wild fires in October, 2007. He received his B.S. in marketing management from the University of Southern California.

Committee Meeting Agendas

MEETING 1: APRIL 28–29, 2009

DAY ONE

8:00–5:00 **Closed Session** (Committee and NRC Staff Only)

DAY TWO

8:00 **Welcome and Working Breakfast**
William Hooke, Chair

8:30 **Request to the National Research Council**
- Why the agency is interested in the problem
- What the agency wants from the study and what it doesn't want
- How the report will be used
- Audience for the report

Report from Department of Homeland Security (DHS)
Michael Dunaway, DHS

10:15 **End of Open Session**

10:15–5:00 **Closed Session** (Committee and NRC Staff Only)

MEETING 2: SEPTEMBER 9–11, 2009
WORKSHOP ON PRIVATE–PUBLIC SECTOR COLLABORATION TO ENHANCE COMMUNITY DISASTER RESILIENCE

DAY ONE

8:30 **Welcome and Introductory Remarks**
William Hooke, Chair

8:45–2:45 **Plenary Session**

Panel One
Why a Collaborative Approach to Community Disaster Resilience Must Become a National Priority

8:45 **Reactions and Reflections**
Panelists: Jason McNamara, Chief of Staff, Federal Emergency Management Agency
Mary Wong, President, Office Depot Foundation
Jim Mullen, Director, Washington State Emergency Management Division

Moderator: Randolph Rowel, Committee Member

9:30 Discussion

Panel Two
Building Community Disaster Resilience through Private–Public Collaboration: What Does it Take to Create and Sustain Effective Cross-sector Partnerships at the State and Local Levels?

10:30 **Best Practices for Establishing Sustainable Partnerships**
Panelists: Brit Weber, Program Director, Michigan State University
Jami Haberl, Executive Director, Safeguard Iowa Partnership
Maria Vorel, National Cadre Manager, Federal Emergency Management Agency

Moderator: Inés Pearce, Committee Member

11:15 Discussion

12:00 **Lunchtime Presentation: The Critical Importance of Community and Cross-Sector Partnerships**
Arif Alikhan, Assistant Secretary for Policy Development, DHS

Panel Three
Making the Business Case: Mobilizing Business to Help Ensure Community and National Disaster Resilience

1:00 **Sustaining Business Involvement in Business-Government Collaboration**
Panelists: Mickie Valente, President, Valente Strategic Advisers, LLC
Stephen Jordan, Executive Director, U.S. Chamber of Commerce
Gene Matthews, Senior Fellow, University of North Carolina

Moderator: Lynne Kidder, Committee Member

1:45 Discussion

2:45–4:30 **Concurrent Sessions**
Workshop participants to break into four groups; each group to discuss both topics.

Factors that facilitate or provide barriers to effective private-public partnerships

- Topic 1: Facilitating Factors
- Topic 2: Barriers

4:30–5:30 **Plenary Session**

4:30 **Summary and Discussion of Concurrent Sessions**

5:30 **Adjourn**

DAY TWO

8:30–4:30 **Plenary Session**

Panel Four
Roles and Perspectives of State and Local Government in Building Community Resilience

8:30 **Fitting in a National Framework**
Panelists: Governor Scott McCallum (Wisconsin, 2001-2003), President and CEO, The Aidmatrix Foundation, Inc.
Ron Carlee, County Manager, Arlington County, Virginia
Leslie Luke, Group Program Manager, County of San Diego, California

Moderator: Michael Lesnick, Committee Member

9:15 Discussion

10:15 **Presentation: The DHS Voluntary Private Sector Preparedness Accreditation and Certification Program**
Emily Walker, National Commission on Terrorist Attacks Upon the United States

11:00 Discussion

11:30 Lunch

12:30–2:00 **Concurrent Sessions**
Workshop participants to break into four groups; each group to discuss both topics.

Building Sustainable Partnerships
- Topic 3: Sustainability
- Topic 4: Resilience-building Efforts and Widespread Implementation

2:15–4:30 **Final Plenary Session**

2:15 **Summary and Discussion of Concurrent Sessions**

3:15 **Presentation: Overarching Workshop Themes**
Brent Woodworth, Committee Member

3:40 **Discussion: Game Changing Ideas and the Path Forward**

4:20 **Closing Remarks**
William Hooke, Chair

4:30 **Adjourn**

DAY THREE

8:30–4:30 **Closed Session** (Committee and NRC Staff Only)

MEETING 3: OCTOBER 19–20, 2009

DAY ONE

8:00 **Welcome and Introductions**
William Hooke, Chair

8:20 **Panel One: Members of the ICMA**
Panelists: Craig Malin, City Manager, Davenport, Iowa
Joyce Wilson, City Manager, El Paso, Texas

Moderator: Lynne Kidder, Committee Member

8:50 Panel Discussion and Question/Answer Period

9:40 **Break**

10:00 **Panel Two: Models for Successful Community Organizing**
Panelists: Claudia Albano, Assistant Public Safety Coordinator, City of Oakland, California
Darius A. Stanton, Vice President, Boys & Girls Clubs of Metropolitan Baltimore, Maryland

Moderator: Arrietta Chakos, Committee Member

10:30 Panel Discussion and Question/Answer Period

11:45 **Discussions Continue during Working Lunch**

1:00 **End of Open Session**

1:00–5:00 **Closed Session** (Committee and NRC Staff Only)

DAY TWO

8:00–5:00 **Closed Session** (Committee and NRC Staff Only)

MEETING 4: DECEMBER 3–4, 2009

Held entirely in closed session (Committee and NRC Staff Only)